Undergraduate Lecture Notes in Physics

Undergraduate Lecture Notes in Physics (ULNP) publishes authoritative texts covering topics throughout pure and applied physics. Each title in the series is suitable as a basis for undergraduate instruction, typically containing practice problems, worked examples, chapter summaries, and suggestions for further reading.

ULNP titles must provide at least one of the following:

- An exceptionally clear and concise treatment of a standard undergraduate subject.
- A solid undergraduate-level introduction to a graduate, advanced, or non-standard subject.
- A novel perspective or an unusual approach to teaching a subject.

ULNP especially encourages new, original, and idiosyncratic approaches to physics teaching at the undergraduate level.

The purpose of ULNP is to provide intriguing, absorbing books that will continue to be the reader's preferred reference throughout their academic career.

More information about this series at http://www.springer.com/series/8917

Wolfram Schmidt · Marcel Völschow

Numerical Python in Astronomy and Astrophysics

A Practical Guide to Astrophysical Problem Solving

 Springer

Wolfram Schmidt
Hamburg Observatory
University of Hamburg
Hamburg, Germany

Marcel Völschow
Hamburg University of Applied Sciences
Hamburg, Germany

ISSN 2192-4791 ISSN 2192-4805 (electronic)
Undergraduate Lecture Notes in Physics
ISBN 978-3-030-70346-2 ISBN 978-3-030-70347-9 (eBook)
https://doi.org/10.1007/978-3-030-70347-9

This Springer imprint is published by the registered company Springer Nature Switzerland AG
The registered company address is: Gewerbestrasse 11, 6330 Cham, Switzerland

Preface

Over the last decades the work of astronomers and astrophysicists has undergone great changes. Making observations is an essential part of astronomy, but most researchers do not operate instruments directly any longer. Most of the time they receive huge amounts of data from remote or even space-bound telescopes and make heavy use of computing power to filter, process, and analyse these data. This requires sophisticated algorithms and, these days, increasingly utilizes machine learning. On the theoretical side of astrophysics, making important discoveries just with pencil and paper belongs to the past (with the occasional exception from the rule). Scientific questions in contemporary astrophysics are often too complex to allow for analytic solutions. As a consequence, numerical computations with a great wealth of physical details play a major role in research now. Back-of-the-envelope calculations still serve their purpose to guide researchers, but at the end of the day it is hardly possible to make progress without writing and running code on computers to gain a deeper understanding of the physical processes behind observed phenomena.

In this regard, it is surprising that the education of students at the undergraduate level is still largely focused on traditional ways of problem solving. It is often argued that being able to program comes along the way, for example, when students engage with their research project for a Bachelor's thesis. It is said that problems in introductory courses can be solved mostly with analytic techniques, and there is no need to bother students with programming languages. However, we are convinced that there is a great deal of computer-based problem solving that can be done right from the beginning. As a result, connections to contemporary science can be made earlier and more lively. One of the obvious merits of becoming acquainted with a programming language is that you can learn how to address a question by developing and implementing an algorithm that provides the answer.

There are two major avenues toward learning a programming language. One follows the systematic teaching model, where the language is laid out in all details and you are guided step by step through its elements and concepts. Surely, this is the preferable method if you want to master a programming language. For the beginner, however, this can become tiresome and confusing, especially since the relevance of most of the stuff you learn becomes clear only later (if at all). The alternative approach is to learn mainly from examples, to grasp the language in an intuitive way and to gradually pick up what you need to know for practical applications. We believe that Python is quite suitable for this approach. Of course, there is always a downside. This textbook is far from covering everything there is to know about Python. We focus on numerical computation and data analysis and make use of libraries geared toward these applications.

Problem solving is an art that requires a lot of practice. The worked-out examples in this book revolve around basic concepts and problems encountered in undergraduate courses introducing astronomy and astrophysics. The complete source code is provided on the web via uhh.de/phy-hs-pybook. We briefly recapitulate essential formulas and basic knowledge, but our recaps are by no means intended to replace lecture courses and textbooks on astronomy and astrophysics. This is highlighted by frequently referring to introductory textbooks for further reading. Our book is mainly intended for readers who want to learn Python from scratch. In the beginning, code examples are explained in detail, and exercises start at a rather elementary level. As topics become more advanced, you are invited to work on problems that require a certain amount of effort, time, and innovative thinking. If you have already experience with programming and know some Python, you can concentrate on topics you are interested in. Our objective is that examples as well as exercises not only help you in understanding and using Python but also offer intriguing applications in astronomy and astrophysics.

Hamburg, Germany   Wolfram Schmidt
December 2020   Marcel Völschow

Acknowledgements

This book was inspired by introductory courses on astronomy and astrophysics at the University of Hamburg. We incorporated Python on the fly into problem classes accompanying the lectures. We thank Robi Banerjee, Jochen Liske, and Francesco de Gasperin for supporting this learning concept, and we are very grateful for the enthusiasm and feedback of many students. Special thanks goes to Bastian R. Brückner, Henrik Edler, Philipp Grete, and Caroline Heneka for reading and commenting the manuscript. Moreover, we thank Bastian R. Brückner for drawing numerous illustrative figures that are invaluable for explaining important concepts in the book.

Contents

Chapter 1
Python Basics

Abstract This chapter explains basic programming concepts. After an overview of common Python distributions, we show how to use Python as a simple calculator. As a first step toward programming, variables and expressions are introduced. The arithmetic series and Fibonacci numbers illustrate the concepts of iteration and branching. We conclude this chapter with a program for the computation of a planet's orbital velocity around the Sun, using constants and functions from libraries and giving a small glimpse at objects in Python.

1.1 Using Python

There is quite a variety of Python installations. Depending on the operating system of your computer, you might have some basic Python preinstalled. Typically, this is the case on Linux computers. However, you might find it rather cumbersome to use, especially if you are not well experienced with writing source code in elementary text editors and executing the code on the command line. What is more, installing additional packages typically requires administrative privileges. If you work, for example, in a computer lab it is likely that you do not have the necessary access rights. Apart from that, Python version 2.x (that is major version 2 with some subversion x) is still in use, while this book is based on version 3.x.

Especially as a beginner, you will probably find it convenient to work with a a GUI (graphical user interface). Two popular choices for Python programming are Spyder and Jupyter. Spyder (www.spyder-ide.org) is a classical IDE (integrated development environment) which allows you to edit code, execute it and view the output in different frames. Jupyter (jupyter.org) can be operated via an arbitrary web browser. It allows you to run an interactive Python session with input and output cells (basically, just like the console-based `ipython`). Apart from input cells for typing Python source code, there are so-called markdown cells for writing headers and explanatory text. This allows you to use formatting similar to elementary HTML for webpages. A valuable feature is the incorporation of LaTeX to display mathematical expressions. The examples in this book can be downloaded as Jupyter notebooks and Python source code in zipped archives from uhh.de/phy-hs-pybook.

© Springer Nature Switzerland AG 2021
W. Schmidt and M. Völschow, *Numerical Python in Astronomy and Astrophysics*,
Undergraduate Lecture Notes in Physics,
https://doi.org/10.1007/978-3-030-70347-9_1

Since it depends on your personal preferences which software suits you best, we do not presume a particular GUI or Python distribution here. If you choose to work with Spyder or Jupyter, online documentation and tutorials will help you to install the software and to get started (browse the official documentation under docs.spyder-ide.org and jupyter-notebook.readthedocs.io/en/stable). For a comprehensive guideline, see also [1, appendices A and B]. A powerful all-in-one solution is Anaconda, a Python distribution and package manager that can be installed under Windows, macOS or Linux by any user (see docs.anaconda.com for more information). Anaconda provides a largely autonomous environment with all required components and libraries on a per-user basis. Of course, this comes at the cost of large resource consumption (in particular, watch your available disk space).

As a first step, check if you can run the traditional "Hello, World!" example with your favorite Python installation. Being astronomers, we use a slightly modified version:

```
1  print("Hello, Universe!")
```

In this book Python source code is listed in frames with lines numbered on the left (in the above example, there is just one line). Although these line numbers are not part of the source code (that's why they are shown outside of the frame), they are useful for referring to particular parts of a code example. You might be able to display line numbers in your code editor (in Jupyter notebooks, for example, line numbering can be switched on and off in the View menu), but you should not confuse these numbers with the line numbers used in this book. Usually we will continue the numbering over several frames if the displayed pieces of code are related to each other, but we also frequently reset line numbers to 1 when a new program starts or a new idea is introduced. Whenever you encounter a code line with number 1 it should alert you: at this point something new begins.

After executing the print statement in line 1 above, you should see somewhere on your screen the output[1]

```
Hello, Universe!
```

The quotes in the source code are not shown in the output. They are used to signify that the enclosed characters form a string. As you might have guessed, the `print`() function puts the string specified in parentheses on the screen (more precisely, in a window or frame that is used by Python for output).[2]

---

[1] How to execute Python code depends on the software you are using (consult the documentation). In a notebook, for example, all you need to do is to simultaneously press the enter and shift keys of your keyboard in the cell containing the code.

[2] Enclosing the string in parentheses is obligatory in Python 3. You may find versions of "Hello, World!" without parentheses on the web, which work only with Python 2.

1.2   Understanding Expressions and Assignments

Apart from printing messages on the screen, which is not particularly exciting by itself, Python can be used as a scientific calculator. Let us begin right away with an example from astronomy. Suppose we want to calculate the velocity at which Earth is moving along its orbit around the Sun. For simplicity, we treat the orbit as circular (in fact, it is elliptical with a small eccentricity of 0.017). From the laws of circular motion it follows that we can simply calculate the velocity as the circumference $2\pi r$ of the orbit divided by the period P, which happens to be one year for Earth. After having looked up the value of π, the orbital radius r (i.e. the distance to the Sun) in km, and the length of a year in seconds,[3] we type

```
1  2*3.14159*1.496e8/3.156e7
```

and, once evaluated by Python, we obtain

```
   29.783388086185045
```

for the orbital velocity in km/s. Line 1 is an example for a Python expression consisting of literal numbers and the arithmetic operators * and / for multiplication and division, respectively. The factor of two in the formula for the circumference is simply written as the integer 2, while the number π is approximately expressed in fixed-point decimal notation as 3.14159.[4] The radius $r = 1.496 \times 10^8$ km is expressed as 1.496e8, which is a so-called floating point literal . The character e followed by an integer indicates the exponent of the leading digit in the decimal system. In this case, e8 corresponds to the factor 10^8. Negative exponents are indicated by a minus sign after e. For example, 10^{-3} can be expressed as 1.0e-3 or just 1e-3 (inserting + for positive exponents is optional).

Of course, there is much more to Python than evaluating literal expressions like a calculator. To get an idea how this works, we turn the example shown above into a little Python program:

```
1  radius = 1.496e8 # orbital radius in km
2  period = 3.156e7 # orbital period in s
3
4  # calculate orbital velocity
5  velocity = 2*3.14159*radius/period
```

Lines 1, 2, and 5 are examples for assignments. Each assignment binds the value of the expression on the right-hand side of the equality sign = to a name on the left-hand side. A value with a name that can be used to refer to that value is in essence what is

---

[3]Strictly speaking, the time needed by Earth to complete one revolution around the Sun is the *sidereal year*, which has about 365.256 d. One day has 86400 s.

[4]In many programming languages, integers such as 2 are treated differently than floating point numbers. For example, using 2.0 instead of the integer 2 in a division might produce a different result. In Python 3, it is usually not necessary to make this distinction. Alas, Python 2 behaves differently in this respect.

called a variable in Python.[5] In line 5, the variables `radius` and `period` are used to compute the orbital velocity of Earth from the formula

$$v = \frac{2\pi r}{P} \tag{1.1}$$

and the result is in turn assigned to the variable `velocity`. Any text between the hash character # and the end of a line is not Python code but a comment explaining the code to someone other than the programmer (once in a while, however, even programmers might be grateful for being reminded in comments about code details in long and complex programs). For example, the comments in line 1 and 2 provide some information about the physical meaning (radius and period of an orbit) and specify the units that are used (km and s). Line 4 comments on what is going on in the following line.

Now, if you execute the code listed above, you might find it surprising that there is no output whatsoever. Actually, the orbital velocity is computed by Python, but assignments do not produce output. To see the value of the velocity, we can append the print statement

```
6  print(velocity)
```

to the program (since this line depends on code lines 1–5 above, we continue the numbering and will keep doing so until an entirely new program starts), which results in the output[6]

```
29.783388086185045
```

This is the same value we obtained with the calculator example at the beginning of this section.

However, we can do a lot better than that by using further Python features. First of all, it is good practice to print the result of a program much in the same way as, hopefully, you would write the result of a handwritten calculation: It should be stated that the result is a velocity in units of km/s. This can be achieved quite easily by using string literals as in the very first example in Sect. 1.1:

```
7  print("orbital velocity =", velocity, "km/s")
```

producing the output

```
orbital velocity = 29.783388086185045 km/s
```

---

[5]The concept of a variable in Python is different from variables in programming languages such as C, where variables have a fixed data type and can be declared without assigning a value. Basically, a variable in C is a placeholder in memory whose size is determined by the data type. Python variables are objects that are much more versatile.

[6]In interactive Python, just writing the variable name in the final line of a cell would also result in its value being displayed in the output.

It is important to distinguish between the word 'velocity' appearing in the string `"orbital velocity ="` on the one hand and the variable `velocity` separated by commas on the other hand. In the output produced by **print** the two strings are concatenated with the value of the variable. Using such a print statement may seem overly complicated because we know, of course, that the program computes the orbital velocity and, since the radius is given in km and the period in seconds, the resulting velocity will be in units of km/s. However, the meaning of a numerical value without any additional information might not be obvious at all in complex, real-world programs producing a multitude of results. For this reason, we shall adhere to the practice of precisely outputting results throughout this book. The simpler version shown in line 6 may come in useful if a program does not work as expected and you want to check intermediate results.

An important issue in numerical computations is the precision of the result. By default, Python displays a floating point value with machine precision (i.e. the maximal precision that is supported by the way a computer stores numbers in memory and performs operations on them). However, not all of the digits are necessarily significant. In our example, we computed the orbital velocity from parameters (the radius and the period) with only four significant digits, corresponding to a relative error of the order 10^{-4}. Although Python performs arithmetical operations with machine precision, the inaccuracy of our data introduces a much larger error. Consequently, it is pointless to display the result with machine precision. Insignificant digits can be discarded in the output by appropriately formatting the value:

```
8 | print("orbital velocity = {:5.2f} km/s".format(velocity))
```

Let us see how this works:

- The method **format** () inserts the value of the variable in parentheses into the preceding string (mind the dot in between). You will learn more about methods in Sect. 1.4.
- The placeholder `{:5.2f}` controls where and in which format the value of the variable `velocity` is inserted. The format specifier `5.2f` after the colon `:` indicates that the value is to be displayed in fixed-point notation with 5 digits altogether (including the decimal point) and 2 digits after the decimal point. The colon before the format specifier is actually not superfluous. It is needed if several variables are formatted in one print statement (examples will follow later).

Indeed, the output now reads

```
orbital velocity = 29.78 km/s
```

Optionally, the total number of digits in the formatting command can be omitted. Python will then just fill in the leading digits before the decimal point (try it; also change the figures in the command and try to understand what happens). A fixed number of digits can be useful, for example, when printing tabulated data.

As the term variable indicates, the value of a variable can be changed in subsequent lines of the program by assigning a new value. For example, you might want to calculate the orbital velocity of a hypothetical planet at ten times the distance of the Earth from the Sun, i.e. $r = 1.496 \times 10^9$ km. To that end, we could start with the assignment `radius=1.496e9`. Alternatively, we can make use of the of the current value based on the assignment in line 1 and do the following:

```
 9   radius = 10*radius
10   print("new orbital radius = {:.3e} km".format(radius))
```

Although an assignment may appear as the equivalent of a mathematical equality, it is of crucial importance to understand that it is not. Transcribing line 9 into the algebraic equation $r = 10r$ is nonsense because one would obtain $1 = 10$ after dividing through r, which is obviously a contradiction. Keep in mind:

> The assignment operator = in Python means *set to*, not *is equal to*.

The code in line 9 thus encompasses three steps:

(a) Take the value currently assigned to `radius`,
(b) multiply this value by ten
(c) and reassign the result to `radius`.

Checking this with the print statement in line 10, we find that the new value of the variable `radius` is indeed 10 times larger than the original value from line 1:

```
new orbital radius = 1.496e+09 km
```

The radius is displayed in exponential notation with three digits after the decimal point, which is enabled by the formatting type e in place of f in the placeholder `{:.3e}` for the radius (check what happens if you use type f in line 10). You must also be aware that repeatedly executing the assignment `radius = 10*radius` in interactive Python increases the radius again and again by a factor of 10, which is possibly not what you might want. However, repeated operation on the same variable is done on purpose in iterative constructions called loops (see Sect. 1.3).

After having defined a new radius, it would not be correct to go straight to the computation of the orbital velocity since the period of the orbit changes, too. The relation between period and radius is given by Kepler's third law of planetary motion, which will be covered in more detail in Sect. 2.2. For a planet in a circular orbit around the Sun, this relation can be expressed as[7]

$$P^2 = \frac{4\pi^2}{GM} r^3, \tag{1.2}$$

---

[7]Here it is assumed that the mass of the planet is negligible compared to the mass of the Sun. For the general formulation of Kepler's third law see Sect. 2.2.

where $M=1.989 \times 10^{30}$ kg is the mass of the Sun and $G=6.674 \times 10^{-11}$ N kg^{-2} m^2 is the gravitational constant. To calculate P for given r, we rewrite this equation in the form

$$P = 2\pi \, (GM)^{-1/2} \, r^{3/2}.$$

This formula can be easily turned into Python code by using the exponentiation operator $**$ for calculating the power of an expression:

```
11  # calculate period in s from radius in km (Kepler's third law)
12  period = 2*3.14159 * (6.674e-11*1.989e30)**(-1/2) * \
13              (1e3*radius)**(3/2)
14  # print period in yr
15  print("new orbital period = {:.1f} yr".format(period/3.156e7))
16
17  velocity = 2*3.14159*radius/period
18  print("new orbital velocity = {:.2f} km/s".format(velocity))
```

The results are

```
new orbital period = 31.6 yr
new orbital velocity = 9.42 km/s
```

Hence, it would take more than thirty years for the planet to complete its orbit around the Sun, as its orbital velocity is only about one third of Earth's velocity. Actually, these parameters are quite close to those of the planet Saturn in the solar system. The backslash character \ in line 12 is used to continue an expression that does not fit into a single line in the following line (there is no limitation on the length of a line in Python, but code can become cumbersome to read if too much is squeezed into a single line). An important lesson taught by the code listed above is that you always need to be aware of physical units when performing numerical calculations. Since the radius is specified in km, we obtain the orbital velocity in km/s. However, the mass of the Sun and the gravitational constants in the expression for the orbital period in lines 12–13 are defined in SI units. For the units to be compatible, we need to convert the radius from km to m. This is the reason for the factor 10^3 in the expression (1e3*radius)**(3/2). Of course, this does not change the value of the variable radius itself. To avoid confusion, it is stated in the comment in line 11 which units are assumed. Another unit conversion is applied when the resulting period is printed in units of a year in line 15, where the expression period/3.156e7 is evaluated and inserted into the output string via format. As you may recall from the beginning of this section, a year has 3.156×10^7 s.

Wrong unit conversion is a common source of error, which may have severe consequences. A famous example is the loss of NASA's Mars Climate Orbiter due to the inconsistent use of metric and imperial units in the software of the spacecraft.[8] As a result, more than \$100 million were quite literally burned on Mars. It is therefore extremely important to be clear about the units of all physical quantities in a

---

[8] See mars.jpl.nasa.gov/msp98/orbiter/.

program. Apart from the simple, but hardly foolproof approach of using explicit conversion factors and indicating units in comments, you will learn different strategies for ensuring the consistency of units in this book.

1.3   Control Structures

The computation of the orbital velocity of Earth in the previous section is a very simple example for the implementation of a numerical algorithm in Python.[9] It involves the following steps:

1. Initialisation of all data needed to perform the following computation.
2. An exactly defined sequence of computational rules (usually based on mathematical formulas), unambiguously producing a result in a finite number of steps given the input from step 1.
3. Output of the result.

In our example, the definition of the variables `radius` and `period` provides the input, the expression for the orbital velocity is a computational rule, and the result assigned to the variable `velocity` is printed as output.

A common generalization of this simple scheme is the repeated execution of the same computational rule in a sequence of steps, where the outcome of one step is used as input for the next step. This is called iteration and will be explained in the remainder of this section. The independent application of the same operations to multiple elements of data is important when working with arrays, which will be introduced in Chap. 2.

Iteration requires a control structure for repeating the execution of a block of statements a given number of times or until a certain condition is met and the iteration terminates. Such a structure is called a loop. For example, let us consider the problem of summing up the first 100 natural numbers (this is a special case of an arithmetic series, in which each term differs by the previous one by a constant):

$$s_n \equiv \sum_{k=1}^{n} k = 1 + 2 + 3 + \ldots + n. \tag{1.3}$$

---

[9]The term algorithm derives from the astronomer and mathematician al-Khwarizmi whose name was transcribed to *Algoritmi* in Latin (cf. [2] if you are interested in the historical background). al-Khwarizmi worked at the *House of Wisdom*, a famous library in Bagdad in the early 9th century. Not only was he the founder of the branch of mathematics that became later known as algebra, he also introduced the decimal system including the digit 0 in a book which was preserved until the modern era only in a Latin translation under the title *Algoritmi de numero Indorum* (this refers to the origin of the number zero in India). The digit 0 is quintessential to the binary system used on all modern computers.

Fig. 1.1 Illustration of the computation of the sum s_{100} defined by Eq. (1.3) via a `for` loop. The box on the top of the figure contains the initialization statement prior to the loop (sum is set to zero). The middle box shows iterations of sum in the body of the `for` loop with the counter k, resulting in the sum shown at the bottom. The arrows indicate how values from one iteration are used in the next iteration. Values of the loop counter are shown in red

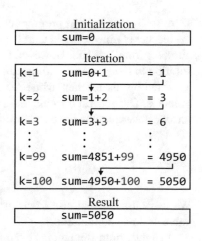

where $n = 100$. In Python, we can perform the summation using a `for` loop:

```
1  sum = 0   # initialization
2  n = 100 # number of iterations
3
4  for k in range(1,n+1):  # k running from 1 to n
5      sum = sum + k       # iteration of sum
6
7  print("Sum =", sum)
```

The result is

```
Sum = 5050
```

The keywords `for` and `in` indicate that the loop counter k runs through all integers defined by `range(1,n+1)`, which means the sequence $1, 2, 3, \ldots, n$ in mathematical notation. It is a potential source of confusion that Python includes the start value 1, but excludes the stop value n+1 in `range(1,n+1)`.[10]

The *indented* block of code following the colon is executed subsequently for each value of the loop counter. You need to be very careful about indentations in Python! They must be identical for all statements in a block, i.e. you are not allowed to use more or less white space or mix tabs and white spaces. We recommend to use one tab per indentation. The block ends with the first non-indented statement. In the example above, only line 5 is indented, so this line constitutes the body of the loop, which adds the value of the loop counter to the variable sum. The initial value of sum must be defined prior to the loop (see line 1). Figure 1.1 illustrates how the variables are iterated in the loop.

---

[10]There is a reason for the stop value being excluded. The default start value is 0 and `range(n)` simply spans the n integers $0, 1, 2, \ldots, n - 1$.

Actually, our Python program computes the sum exactly in the way intended by the teacher of nine-year old Carl Friedrich Gauss[11] in school, just by summing up the numbers from 1 to 100. However, Gauss realized that there is a completely different solution to the problem and he came up with the correct answer much faster than expected by his teacher, while his fellow students were still tediously adding up numbers. The general formula discovered by Gauss is (the proof can be found in any introductory calculus textbook or on the web):

$$s_n = \sum_{k=1}^{n} k = \frac{n(n-1)}{2} . \tag{1.4}$$

We leave it as an exercise to check with Python that this formula yields the same value as direct summation.

A slightly more demanding example is the calculation of the Fibonacci sequence[12] using the recursion formula

$$F_{n+1} = F_n + F_{n-1} \quad \text{for } n \geq 1, \tag{1.5}$$

with the first two elements

$$F_1 = 1, \quad F_0 = 0 . \tag{1.6}$$

The meaning of Eq. (1.5) is that any Fibonacci number is the sum of the two preceding ones, starting from 0 and 1. The following Python program computes and prints the Fibonacci numbers $F_1, F_2, \ldots, F_{10}$ (or as many as you like):

```python
1   # how many numbers are computed
2   n_max = 10
3
4   # initialize variables
5   F_prev = 0 # 0. number
6   F = 1       # 1. number
7
8   # compute sequence of Fibonacci numbers
9   for n in range(1,n_max+1):
10      print("{:d}. Fibonacci number = {:d}".format(n,F))
11
12      # next number is sum of F and the previous number
13      F_next = F + F_prev
14
```

---

[11]German mathematician, physicist, and astronomer who is known for the Gauss theorem, the normal distribution, and many other import contributions to algebra, number theory, and geometry.

[12]Named after Leonardo de Pisa, also known as Fibonacci, who introduced the sequence to European mathematics in the early 13th century. However, the Fibonacci sequence was already known to ancient Greeks. It was used to describe growth processes and there is a remarkable relation to the golden ratio.

```
15      # prepare next iteration
16      F_prev = F # first reset F_prev
17      F = F_next # then assign next number to F
```

The three variables F_prev, F, and F_next correspond to F_{n-1}, F_n, and F_{n+1}, respectively. The sequence is initialized in lines 5 and 6 (definition of F_0 and F_1). The recursion formula (1.5) is implemented in line 13. Without lines 16 and 17, however, the same value (corresponding to $F_0 + F_1$) would be assigned again and again to F_next. For the next iteration, we need to re-assign the values of F (F_n) and F_next (F_{n+1}) to F_prev (F_{n-1}) and F (F_n), respectively. The loop counter n merely controls how many iterations are executed. Figure 1.2 shows a schematic view of the algorithm (you can follow the values of the variables in the course of the iteration by inserting a simple print statement into the loop). The output produced by the program is[13]

```
 1. Fibonacci number = 1
 2. Fibonacci number = 1
 3. Fibonacci number = 2
 4. Fibonacci number = 3
 5. Fibonacci number = 5
 6. Fibonacci number = 8
 7. Fibonacci number = 13
 8. Fibonacci number = 21
 9. Fibonacci number = 34
10. Fibonacci number = 55
```

Since two variables are printed, we need two format fields where the values of the variables are inserted (see line 10). As an alternative to using **format** (), the same output can be produced by means of a formatted string literal (also called f-string)[14]:

```
print(f"{n:d}. Fibonacci number = {F:d}")
```

Here, the variable names are put directly into the string. The curly braces indicate that the values assigned to the names n and F are to be inserted in the format defined after the colons (in this example, as integers with arbitrary number of digits). This is a convenient shorthand notation. Nevertheless, we mostly use **format** () in this book because the syntax maintains a clear distinction between variables and expressions on the one hand and formatted strings on the other hand. If you are more inclined to f strings, make use of them as you please.

Suppose we would like to know all Fibonacci numbers smaller than, say, 1000. We can formally write this as $F_n < 1000$. Since it is not obvious how many Fibonacci numbers exist in this range, we need a control structure that repeats a block of code

---

[13] As you can see from Fig. 1.2, F_{11} is computed as final value of F_next. But it is not used. You can try to modify the program such that only 9 iterations are needed to print the Fibonacci sequence up to F_{10}.

[14] This feature was introduced with Python 3.6.

Fig. 1.2 Illustration of the recursive computation of the Fibonacci sequence (see Eq. 1.5). In each iteration of the loop, the sum of `F_prev` and `F` is assigned to the variable `F_prev`. The resulting number is shown in the rightmost column. For the next iteration, this number is re-assigned to `F`, and the value of `F` to `F_prev`, as indicated by the arrows

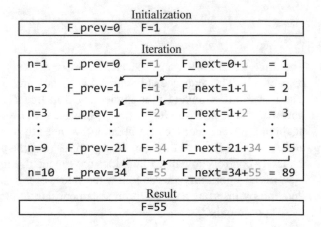

as long as a certain condition is fulfilled. In such a case, it is preferable to work with a `while` loop. This type of loop enables us to modify our program such that all Fibonacci numbers smaller than 1000 are computed, without knowing the required number of iterations:

```
1   # initialize variables
2   F_prev = 0 # 0. number
3   n,F = 1,1   # 1. number
4
5   # compute sequence of Fibonacci numbers smaller than 1000
6   while F < 1000:
7       print("{:d}. Fibonacci number = {:d}".format(n,F))
8
9       # next number is sum of F and the previous number
10      F_next = F + F_prev
11
12      # prepare next iteration
13      F_prev = F # first reset F_prev
14      F = F_next # then assign next number to F
15      n += 1      # increment counter
```

The resulting numbers are:

```
1. Fibonacci number = 1
2. Fibonacci number = 1
3. Fibonacci number = 2
4. Fibonacci number = 3
5. Fibonacci number = 5
6. Fibonacci number = 8
7. Fibonacci number = 13
```

```
 8. Fibonacci number = 21
 9. Fibonacci number = 34
10. Fibonacci number = 55
11. Fibonacci number = 89
12. Fibonacci number = 144
13. Fibonacci number = 233
14. Fibonacci number = 377
15. Fibonacci number = 610
16. Fibonacci number = 987
```

Of course, the first ten numbers are identical to the numbers from our previous example. If you make changes to a program, always check that you are able to reproduce known results!

The loop header in line 6 of the above listing literally means: Perform the following block of code *while* the value of F is smaller than 1000. The expression F<1000 is an example of a Boolean (or logical) expression. The operator < compares the two operands F and 1000 and evaluates to True if the numerical value of F is smaller than 1000. Otherwise, the expression evaluates to False and the loop terminates. Anything that is either True or False is said to be of Boolean type. In Python, it is possible to define Boolean variables.

A while loop does not come with a counter. To keep track of how many Fibonacci numbers are computed (in other words the index n of the sequence F_n), we initialize the counter n along with F in the multiple assignment in line 3. This is equivalent to

```
n = 1
F = 1
```

Python allows you to assign multiple values separated by commas to multiple variables (also separated by commas) in a single statement, where the ordering on the left corresponds to the ordering on the right. We will make rarely use of this feature. While it is useful in some cases (for example, to swap variables[15] or for functions returning multiple values), multiple assignments are rather difficult to read and prone to errors, particularly if variables are interdependent.

While the loop counter of a for loop is automatically incremented, we need to explicitly increase our counter in the example above at the end of each iteration. In line 15, we use the operator += to increment n in steps of one, which is equivalent to the assignment n=n+1 (there are similar operators -=, *=, etc. for the other basic arithmetic operations)

Let us try to be even smarter and count how many even and odd Fibonacci numbers below a given limit exist. This requires branching, i.e. one block of code will be executed if some condition is met and an alternative block if not (such blocks are also called clauses). This is exactly the meaning of the if and else statements in the following example:

---

[15] Another application in our Fibonacci program would be the merging of lines 13 and 14 into the multiple assignment F_prev, F = F, F_next.

```
1   # initialize variables
2   F_prev = 0 # 0. number
3   F = 1        # 1. number
4   n_even = 0
5   n_odd = 0
6
7   # compute sequence of Fibonacci numbers smaller than 1000
8   while F < 1000:
9       # next number is sum of F and the previous number
10      F_next = F + F_prev
11
12      # prepare next iteration
13      F_prev = F # first reset F_prev
14      F = F_next # then assign next number to F
15
16      # test if F is even (divisible by two) or odd
17      if F%2 == 0:
18          n_even += 1
19      else:
20          n_odd += 1
21
22  print("Found {:d} even and {:d} odd Fibonacci numbers".\
23          format(n_even,n_odd))
```

Instead of a single counter, we need two counters here, n_even for even Fibonacci numbers and n_odd for the odd ones. The problem is to increment n_even if the value of F is an even number. To that end the modulo operator % is applied to get the remainder of division by two. If the remainder is zero, then the number is even. This is tested with the comparison operator == in the Boolean expression following the keyword **if** in line 17. If this expression evaluates to True, then the counter for even numbers is incremented (line 18). If the condition is False, the **else** branch is entered and the counter for odd numbers is incremented (line 20). You must not confuse the operator ==, which *compares* variables or expressions *without changing* them, with the assignment operator =, which sets the value of a variable. The program reports (we do not bother to print the individual numbers again):

```
Found 5 even and 11 odd Fibonacci numbers
```

Altogether, there are $5 + 11 = 16$ numbers. You may check that this is in agreement with the listed numbers.

1.4   Working with Modules and Objects

Python offers a collection of useful tools in the Python Standard Library (see docs.python.org/3/library). Functions such as **print**() are part of the Standard Library. They are called *built-in* functions. Apart from that, many more optional libraries (also called packages) are available. Depending on the Python distribution you use, you will find that some libraries are included and can be imported as shown below, while you might need to install others.[16] Python libraries have a hierarchical modular structure. This means that you do not necessarily have to load a complete library, but you can access some part of a library, which can be a module, a submodule (i.e. a module within a module) or even individual names defined in a module. To get started, it will be sufficient to consider a module as a collection of definitions. By importing a module, you can use variables, functions, and classes (see below) defined in the module.

For example, important physical constants and conversion factors are defined in the `constants` module of the SciPy library (for more information, see www.scipy.org/about.html). A module can be loaded with the **import** command:

```
1  import scipy.constants
```

To view an alphabetically ordered list of all names defined in this module, you can invoke **dir**(`scipy.constants`) (this works only after a module is imported). By scrolling through the list, you might notice the entry `'gravitational_constant'`. As the name suggests, this is the gravitational constant G. Try

```
2  print(scipy.constants.gravitational_constant)
```

which displays the value of G in SI units:

```
6.67408e-11
```

The same value is obtained via `scipy.constants.G`. Even so, an identifier composed of a library name in conjunction with a module and a variable name is rather cumbersome to use in programs. Alternatively, a module can be accessed via an alias:

```
3  import scipy.constants as const
4
5  print(const.G)
```

also displays the value of G. Here, `const` is a user-defined nickname for `scipy.constants`.

It is also possible to import names from a module directly into the global namespace of Python. The variables we have defined so far all belong to the global namespace. The syntax is as follows:

```
6  from scipy.constants import G
7
8  print(G)
```

---

[16]See, for example, packaging.python.org/tutorials/installing-packages
and docs.conda.io/projects/conda/en/latest/user-guide/tasks/manage-pkgs.html.

In this case, only G is imported, while in the examples above *all* names from
`scipy.constants` are made available. Importing names via the keyword `from`
should be used with care, because they can easily conflict with names used in assignments elsewhere. Python does not treat this as an error. Consequently, you might
accidentally overwrite a module variable such as G with some other value.

By using constants from Python libraries, we can perform computations without
looking up physical constants on the web or in textbooks and inserting them as literals
in the code. Let us return to the example of a planet at 10 times the distance of Earth
from the Sun, i.e. $r = 10$ au (see Sect. 1.2). Here is an improved version of the code
for the computation of the orbital period and velocity:

```
 1  from math import pi,sqrt
 2  from astropy.constants import M_sun
 3  from scipy.constants import G,au,year
 4
 5  print("1 au =", au, "m")
 6  print("1 yr =", year, "s")
 7
 8  radius = 10*au
 9  print("\nradial distance = {:.1f} au".format(radius/au))
10
11  # Kepler's third law
12  period = 2*pi * sqrt(radius**3/(G*M_sun.value))
13  print("orbital period = {:.4f} yr".format(period/year))
14
15  velocity = 2*pi * radius/period # velocity in m/s
16  print("orbital velocity = {:.2f} km/s".format(1e-3*velocity))
```

The output of this program is

```
1 au = 149597870691.0 m
1 yr = 31536000.0 s

radial distance = 10.0 au
orbital period = 31.6450 yr
orbital velocity = 9.42 km/s
```

We utilize the value of π and the square-root function defined in the `math` module,
which is part of the standard library. The function `sqrt()` imported from `math` is
called in line 12 with the expression `radius**3/(G*M_sun.value)` as argument. This means that the number resulting from the evaluation of this expression is
passed as input to `sqrt()`, which executes an algorithm to compute the square root
of that number. The result returned by the function and is then multiplied with `2*pi`
to obtain the orbital period. Moreover, we use constants and conversion factors from
the SciPy and Astropy libraries. For instance, `au` is one astronomical unit in m and
`year` is one year in s. The values are printed in lines 5 and 6. These conversion
factors enable us to conveniently define the radius in astronomical units (line 8) and,
after apply Kepler's third law in SI units, to print the resulting orbital period in years
(lines 12 and 13). When printing the radius in line 9, the newline character `'\n'`
at the beginning of the string inserts a blank line. Other than in Sect. 1.2, the orbital

velocity assigned to `velocity` is in m/s. So we need to multiply by a factor of 10^{-3} to obtain the velocity in units of km/s in the final print statement.

In contrast to the gravitational constant G, which is simply a floating point number, the mass of the Sun defined in `astropy.constants` is a more complex object. In computer science, the term object has a specific meaning and refers to the object-oriented programming (OOP) paradigm. You can go a long way in Python without bothering about object-oriented programming. Nevertheless, you will find it helpful if you are aware of a few basic facts:

1. Everything in Python is an object.
2. An object contains a particular kind of data.
3. Objects have methods to manipulate the object's data in a controlled way.
4. A method can change an object or create a new object.

This implies that G is also an object, albeit a rather simple one. If you print `M_sun`, you will find quite a bit more information in there, such as the uncertainty of the value and its physical unit:

```
Name    = Solar mass
Value   = 1.9884754153381438e+30
Uncertainty  = 9.236140093538353e+25
Unit  = kg
Reference = IAU 2015 Resolution B 3 + CODATA 2014
```

Particular data items are called object attributes. For example, the value of the solar mass is an attribute of `M_sun`. You can fetch an attribute by joining the names of the object and the attribute with a dot. We refer to the attribute `value` in line 12 to obtain a pure number that can be combined with other numbers in an arithmetic expression. To list attributes belonging to an object, you can use `dir()`, just like for modules, or search the documentation. Attributes and methods are defined in classes. While objects belonging to the same class may contain different data, they have the same attributes and methods. For example, `M_earth` from `astropy.constants` has a `value` attribute just like `M_sun`, but the value behind this attribute is Earth's mass instead of the mass of the Sun. Both objects belong to the class `Quantity`. You can take a glimpse behind the curtain in Appendix A, where you are briefly introduced to writing your own classes.

Since everything in Python is an object, so is a string. Now you are able to better understand the meaning of `format()` being a method. It is a method allowing you to insert formatted numbers into a string.[17] While methods are relatives of Python functions, a method always has to be called in conjunction with a particular object. In the print statements in lines 9, 13, and 16, the objects are string literals. The

---

[17] Since strings are immutable objects, the method does not change the original string with place-holders. It creates a new string object with the formatted numbers inserted.

syntax is similar to accessing object attributes, except for the arguments enclosed in parentheses (here, the variables holding the numbers to be inserted). In general, you can call methods on names referring to objects – in other words, Python variables. This will be covered in more detail in the next chapter.

Chapter 2
Computing and Displaying Data

Abstract NumPy arrays are the workhorses of numerics in Python, extending it by remarkable numerical capabilities. For example, they can be used just like simple variables to evaluate an arithmetic expression for many different values without programming a loop. In the first section, we combine the power of NumPy and Astropy and compute the positions of objects on the celestial sphere. Moreover, we introduce Matplotlib to produce plots from array data. Further applications are shown in the context of Kepler's laws and tidal forces, for example, printing formatted tables and plotting vector maps.

2.1 Spherical Astronomy

In astronomy, the positions of stars are specified by directions on the sky, i.e. two angular coordinates. The radial distance of the star from Earth would be the third coordinate, but distances are not known for all astronomical objects. As long as distances do not matter, all astronomical objects can be projected in radial directions onto a sphere with Earth at its center (the size of the sphere does not matter, but you can think of it as a distant spherical surface). This is the so-called celestial sphere.[1]

For an observer on Earth, the position of astronomical objects depends on geographical latitude and longitude and varies with the time of day. To specify positions independent of the observer, angular coordinates with respect to fixed reference directions are used. In the equatorial coordinate system, one reference direction is given by Earth's rotation axis (or, equivalently, the orientation of the equatorial plane). The orientation of the rotation axis is fixed because of angular momentum conservation,[2] The angular distance of a star from the equatorial plane is called declination and denoted by δ. The other reference direction is defined by the intersection between

---

[1] In ancient and medieval times, it was thought that there is actually a physical sphere with the stars attached to it and Earth at its center. This is known as geocentric world view. The celestial sphere in modern astronomy is merely a useful mathematical construction with no physical meaning whatsoever.

[2] This is not exactly true since other bodies in the solar system cause perturbations, but changes are sufficiently small over a human's lifetime.

© Springer Nature Switzerland AG 2021
W. Schmidt and M. Völschow, *Numerical Python in Astronomy and Astrophysics*,
Undergraduate Lecture Notes in Physics,
https://doi.org/10.1007/978-3-030-70347-9_2

Fig. 2.1 Celestial sphere with coordinates α (right ascension) and δ (declination) of a stellar object. Earth's orbital plane (ecliptic) intersects the sphere along the red circle, which is inclined by the angle ϵ_0 (obliquity) with respect to the celestial equator. The celestial equator is the outward projection of Earth's equator onto the celestial sphere. The intersection points of the ecliptic and the celestial equator are the two equinoxes

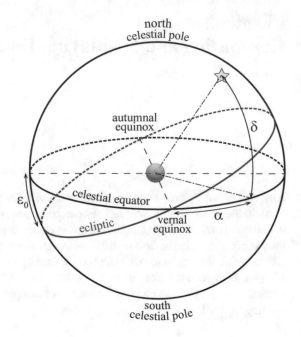

the equatorial plane and the plane of Earth's orbital motion around the Sun. The fixed orientation of the orbital plane, which is called ecliptic, is also a consequence of angular momentum conservation (in this case the angular momentum of orbital motion). The second coordinate in the equatorial system, which is called right ascension α, is the angle measured from one of the two points where the celestial sphere is pierced by the line of intersection of the equatorial an orbital planes. The zero point for the right ascension is known as vernal equinox, the opposite point as autumnal equinox. If all of this sounds rather complicated, it will become clear from the illustration in Fig. 2.1. See also [3, Sect. 2.5].

2.1.1 Declination of the Sun

While the declination of stars is constant, the position of the Sun changes in the equatorial system over the period of a year. This is a consequence of the inclination of Earth's rotation axis with respect to the direction perpendicular to the ecliptic, which is equal to $\epsilon_0 = 23.44°$. The angle ϵ_0 is called obliquity of the ecliptic. The annual variation of the declination of the Sun is approximately given by[3]

$$\delta_\odot = -\arcsin\left[\sin\epsilon_0\,\cos\left(\frac{360°}{365.24}\,(N+10)\right)\right] \tag{2.1}$$

---

[3] See en.wikipedia.org/wiki/Position_of_the_Sun.

where N is the difference in days starting from 1st January. So the first day of the year corresponds to $N = 0$, and the last to $N = 364$ (unless it is a leap year). The fraction $360°/365.24$ equals the change in the angular position of Earth per day, assuming a circular orbit. This is just the angular velocity ω of Earth's orbital motion around the Sun in units of degrees per day.[4]

The Sun has zero declination at the equinoxes (intersection points of celestial equator and ecliptic) and reaches $\pm\epsilon_0$ at the solstices, where the rotation axis of the Earth is inclined towards or away from the Sun. The exact dates vary somewhat from year to year. In 2020, for instance, equinoxes were on 20th March and 22nd September and solstices on 20th June and 21st December (we neglect the exact times in the following). In Exercise 2.2 you are asked to determine the corresponding values of N. For example, the 20th of June is the 172nd day of the year 2020. Counting from zero, we thus expect the maximum of the declination $\delta_\odot = \epsilon_0$ (first solstice) for $N = 171$. Let us see if this is consistent with the approximation (2.1). The following Python code computes the declination $\delta_\odot$ based on this formula for a given value of N:

```python
import math

N = 171 # day of 1st solstice
omega = 2*math.pi/365.24 # angular velocity in rad/day
ecl = math.radians(23.44) # obliquity of the ecliptic

# approximate expression for declination of the Sun
delta = -math.asin(math.sin(ecl)*math.cos(omega*(N+10)))
print("declination = {:.2f} deg".format(math.degrees(delta)))
```

The result

```
declination = 23.43 deg
```

is close to the expected value of $23.44°$. To implement Eq. (2.1), we use the sine, cosine, and arcsine functions from the `math` library. The arguments of these functions must be specified in radians. While the angular velocity is simply given by $2\pi/365.24$ rad/d (see line 4), ϵ_0 is converted into radians with the help of `math.radians()` in line 5. Both values are assigned to variables, which allows us to reuse them in subsequent parts of the program. Since the inverse function `math.asin()` returns an angle in radians, we need to convert `delta` into degrees when printing the result in line 9 ('deg' is short for degrees).

To calculate the declination for the second solstice and also for the equinoxes, we need to evaluate the code in line 8 for the corresponding values of N. If you put the code listed above into a Python script, you can change the value assigned to the variable N and simply re-run the script. While this is certainly doable for a few different values, it would become too tedious for many values (we will get to that soon enough). Ideally, we would like to compute the Sun's declination for several

---

[4] The angular velocity in radians per unit time appears as factor $2\pi/P$ in Eq. (1.1).

days at once. This can be done by using data structures know as arrays. An array is an ordered collection of data elements of the same type. Here is as an example:

```
10   import numpy as np
11
12   # equinoxes and solstices in 2020
13   N = np.array([79, 171, 265, 355])
```

In contrast to other programming languages, arrays are not native to Python. They are defined in the module numpy (the library name is also written as NumPy, see www.numpy.org), which is imported under the alias np in line 10. The function np.array() takes a list of values in brackets and, if possible, creates an array with these values as elements. Like an array, a list is also an ordered collection of data elements. In this book, we will rarely make use of lists (see, for example, Sect. 5.5). They are more flexible than NumPy arrays, but flexibility comes at the cost of efficiency. This is demonstrated in more detail in Appendix B.1. On top of that, NumPy offers a large toolbox of numerical methods which are specifically implemented to work with arrays.[5] The array returned by np.array() is assigned to the variable N (mark the difference between single value versus array in lines 3 and 13, respectively). Its main properties can be displayed by the following print statements:

```
14   print(N)
15   print(N.size)
16   print(N.dtype)
```

which produces the output

```
[ 79 171 265 355]
4
int64
```

From this we see that N has four elements (the number of elements is obtained with the .size attribute), which are the integers 79, 171, 256, and 355. The data type can be displayed with the .dtype attribute (by default, 64-bit integers are used for literals without decimal point).

How can we work with the values in an array? For example, N[0] refers to the first element of the array, which is the number 79. Referring to a single element of an array via an integer in brackets, which specifies the *position* of the element in the array, is called indexing. The first element has index 0, the second element index 1 and so on (remember, this is Python's way of counting). You can also index elements from the end of the array. The last element has index -1, the element before the last one has index -2, etc. So, for the array defined above,

```
17   print(N[1])
18   print(N[-3])
```

---

[5]See also [1] for a comprehensive introduction to arrays.

outputs

```
171
171
```

Do you see why the value 171 is printed in both cases? Vary the indices and see for yourself what you get. Before proceeding, let us summarize the defining properties of an array:

> Every element of an array must have the same data type. Each element is identified by its index.

Having defined N as an array, we could calculate the declination of the Sun on day 171 (the first solstice) by copying the code from line 8 and replacing N by N[1] in the expression for delta. Of course, this would not bring us any closer to calculating the declination of the Sun for all four days *at once*. With NumPy, it can be done as follows.

```
19  delta = -np.arcsin(math.sin(ecl) * np.cos(omega*(N+10)))
20  print(np.degrees(delta))
```

In short, the expression in line 19 is evaluated *for each element* of the array N and the results are stored in a new array that is assigned to delta. The four elements of delta are the values of the declination for the two equinoxes and solstices:

```
[ -0.9055077   23.43035419  -0.41950731 -23.43978827]
```

Not perfect (the declination at equinoxes should be zero), but reasonably close. We use only an approximate formula after all.

To get a better idea of how the code in line 19 works, we expand it into several steps and use a temporary array called tmp for intermediate results of the calculation:

```
21  # add 10 to each element of N
22  tmp = N+10
23  print(tmp)
24  print(tmp.dtype)
25
26  # multipy by omega
27  tmp = omega*tmp
28  print(tmp)
29  print(tmp.dtype)
30
31  # calculate the cosine of each element in the resulting array
32  # and multipy with the sine of the obliquity
33  tmp = math.sin(ecl)*np.cos(tmp)
```

```
34   print(tmp)
35
36   # calculate the negative arcsine of each element
37   delta = -np.arcsin(tmp)
38   print(np.degrees(delta))
```

As you can see, it is possible to perform arithmetic operations on arrays. For operators such as + and *, the operands must be (see also Fig. 2.2)

- either an array and a single number,
- or two arrays of *identical shape* (for arrays with only one index this amounts to the same number of elements).

For example, we make use of the first option in line 22, where the number 10 is added to each element of the array N. The resulting integers are assigned element-wise to the array tmp:

```
[ 89 181 275 365]
int64
```

The next step is an element-wise multiplication with omega (Earth's angular velocity defined in line 4). Since the value of omega is a floating point number, the tmp array is automatically converted from data type integer to float:

```
[1.53105764 3.11372396 4.73079608 6.27905661]
float64
```

Now we have the angular positions of Earth on its orbit, from which the vertical distances of the Sun from the celestial equator can be computed. In line 33, the NumPy function np.cos() is used to compute the cosine of each value in the tmp array, while math.sin() can take only a single-valued argument, which is ecl (the obliquity of the ecliptic). The product is

```
[ 0.01580343 -0.39763404  0.00732172  0.39778512]
```

and, after computing the arcsine of each element with np.arcsin, we finally get the declinations:

```
[ -0.9055077   23.43035419  -0.41950731 -23.43978827]
```

While doing things step by step is helpful for beginners, combining all these steps into the single statement shown in line 19 is extremely useful for the more experienced programmer. Starting with the code for a single day (see line 8), the only modification that has to be made is that the math module has to be replaced by numpy whenever the argument of a function is an array. However, there is small pitfall. By comparing the code examples carefully, you might notice that the identifiers for the arcsine function read math.asin() and np.arcsin() in lines 8 and 19, respectively.

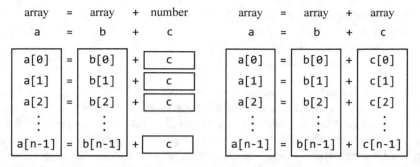

Fig. 2.2 Illustration of basic NumPy operations. Adding a number (variable) to an array, means that the same number is added to each element in the array (left). If two arrays of size n are added, the operator + is applied element-wise (right)

A likely mistake is to replace `math.asin()` by `np.asin()`, which would cause Python to throw the following error message:

```
AttributeError: module 'numpy' has no attribute 'asin'
```

Since `math` and `numpy` are independent modules, you cannot expect that the naming of mathematical functions is always consistent.

Printing an array, results in all elements of the array being displayed in some default format. Now, suppose we want to format the elements nicely as we did at the beginning of this section before introducing arrays. Formatted printing of a particular element of an array is of course possible by indexing the element. For example,

```
39  print("declination = {:.2f} deg".
40       format(math.degrees(delta[1])))
```

produces the same output as line 9 (where `delta` is a simple variable):

```
declination = 23.43 deg
```

To display all elements in a formatted table, we need to loop through the array. Actually, we have already worked with such loops, for example, in line 19. Loops of this type are called implicit. The following example shows an explicit **for** loop:

```
41  for val in delta:
42      print("declination = {:6.2f} deg".
43           format(math.degrees(val)))
```

The loop starts with the first element in the array `delta`, assigns its value to the variable `val`, which is the loop variable (similar to the loop counter introduced in Sect. 1.3), and then executes the loop body. The loop in the example above encompasses only a single print statement. After executing this statement, the loop continues with the next element and so on until the end of the array (the highest index) is reached:

```
declination =   -0.91 deg
declination =   23.43 deg
declination =   -0.42 deg
declination = -23.44 deg
```

However, the output is not really satisfactory yet. While repeatedly printing
`declination` is redundant, important information for understanding the data is
missing. In particular, the days to which these values refer are not specified. A better way of printing related data in arrays is, of course, a table. In order to print the
days and the corresponding declinations, we need to simultaneously iterate through
the elements of N and `delta`. One solution is to define a counter with the help of
Python's **enumerate**() function:

```
44  print("i   day   delta [deg]")
45  for i,val in enumerate(delta):
46      print("{1:d}   {2:3d}   {0:8.2f}".\
47              format(math.degrees(val),i,N[i]))
```

Compared to the **for** loops discussed in Sect. 1.3, the range of the counter i is
implicitly given by the size of an array: It counts through all elements of `delta`
and, at the same time, enables us to reference the elements of N sequentially in the
loop body. Execution of the loop produces the following table:

```
i   day   delta [deg]
0    79      -0.91
1   171      23.43
2   265      -0.42
3   355     -23.44
```

Here, the arguments of **format**() are explicitly ordered. The placeholder {1:d} in
line 46 indicates that the second argument (index 1 before the colon) is to be inserted
as integer at this position. The next placeholder refers to the third argument (index 2
before the colon) and the last one to the *first* argument. The format specifiers ensure
sufficient space between the columns of numbers (experiment with the settings). The
header with labels for the different columns is printed in line 44 (obviously, this has
to be done only once before the loop begins).

Surely, instead of enumerating `delta`, you could just as well enumerate N. Is
it possible to write a loop that simultaneously iterates both arrays without using
an explicit counter? Actually, this is the purpose of the **zip**() function, which
aggregates elements with the same index from two or more arrays of equal length.
In this case, the loop variable is a tuple containing one element from each array.
We will take a closer look at tuples below. All you need to know for the moment is
that the tuple `row` in the following code example contains one element of N and the
corresponding element of `delta`, forming one row of the table we want to print.
Tuple elements are indexed like array elements, so `row[0]` refers to the day and
`row[1]` to the declination for that day.

```
48  print("day  delta [deg]")
49  for row in zip(N,delta):
50      print("{0:3d}  {1:8.2f}".
51                format(row[0],math.degrees(row[1])))
```

Apart from the index column, we obtain the same output as above:

```
day  delta [deg]
 79     -0.91
171     23.43
265     -0.42
355    -23.44
```

If you replace lines 50–51 by the unformatted print statement **print**(row), you will get:

```
(79, -0.01580409076383853)
(171, 0.40893682550286947)
(265, -0.007321783769611206)
(355, -0.4091014813515704)
```

While brackets are used for lists and arrays, tuples are enclosed by parentheses. The most important difference is that tuples are immutable, i.e. it is not possible to add, change or remove elements. For example, the tuples shown above are fixed pairs of numbers, similar to coordinates (x, y) for points in a plane. Suppose you want to change day 355 to 364. While it is possible to modify array N by setting N[3] = 364, you are not allowed to assign a value to an individual element in a tuple, such as row[0]. You can only overwrite the whole tuple by a new one (this is what happens in the loop above).

2.1.2 Diurnal Arc

From the viewpoint of an observer on Earth, the apparent motion of an object on the celestial sphere follows an arc above the horizon, which is called diurnal arc (see Fig. 2.3). The time-dependent horizontal position of the object is measured by its hour angle h. An hour angle of 24^h corresponds to a full circle of 360° parallel to the celestial equator (an example is the complete red circle in Fig. 2.3). For this reason, h is can be equivalently expressed in degrees or radians. However, as we will see below, an hour angle of 1^h is not equivalent to a time difference of one solar hour. By definition the hour angle is zero when the object reaches the highest altitude above the horizon (see also Exercise 2.4 and Sect. 2.1.3). The hour angle corresponding to the setting time, when the object just vanishes beneath the horizon, is given by[6]

---

[6]See [3, Sect. 2.6] if you are interested in how this equation comes about.

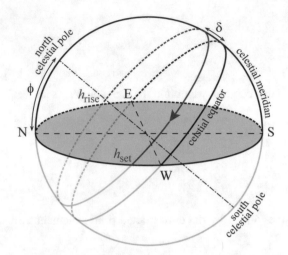

Fig. 2.3 Diurnal arc of a star moving around the celestial sphere (thick red circle) in the horizontal system (see Sect. 2.1.3) of an observer at latitude ϕ (the horizontal plane is shown in grey). Since the equatorial plane is inclined by the angle $90° - \phi$ against the horizontal plane, the upper culmination of the star at the meridian is given by $a_{max} = 90° - \phi + \delta$, where δ is the declination. In the co-rotating system, the star rises at hour angle h_{rise}, reaches its highest altitude when it crosses the meridian at $h = 0$, and sets at the horizon at $h_{set} = -h_{rise}$

$$\cos h_{set} = -\tan \delta \, \tan \phi, \qquad (2.2)$$

where δ is the declination of the object (see Sect. 2.1) and ϕ the latitude of the observer's position on Earth. As a consequence, the variable $T = 2h_{set}$ measures the so-called sidereal time for which the object is in principle visible on the sky (stars are of course outshined by the Sun during daytime). It is also known as length of the diurnal arc.

For example, let us consider the star Betelgeuse in the constellation of Orion. It is a red giant that is among the brightest stars on the sky. Its declination can be readily found with the help of `astropy.coordinates`, which offers a function that searches the name of an object in online databases:

```
1  from astropy.coordinates import SkyCoord, EarthLocation
2
3  betelgeuse = SkyCoord.from_name('Betelgeuse')
4  print(betelgeuse)
```

When you are confronted with the output for the first time, it might require a little bit of deciphering:

```
<SkyCoord (ICRS): (ra, dec) in deg
    (88.79293899, 7.407064)>
```

This tells us that the right ascension (ra) and declination (dec) of the object named `Betelgeuse` were found to be 88.79° and 7.41°, respectively.[7] The variable `betelgeuse` defined in line 3 represents not only an astronomical object; it is a Python object, more specifically an object belonging to the class `SkyCoord` (see Sect. 1.4 for objects in a nutshell). The attribute `dec` allows us to directly reference the declination:

```
5  delta = betelgeuse.dec
6  print(delta)
```

The declination is conveniently printed in degrees (d), arc minutes (m) and arc seconds (s), which is the preferred format to express angular coordinates in astronomy:

```
7d24m25.4304s
```

i.e. $\delta \approx +07° 24' 25''$, where $1' = (1/60)°$ and $1'' = (1/60)'$.

Suppose we want to determine the length of Betelgeuse's diurnal arc as seen from Hamburg Observatory ($\phi \approx +53° 28' 49''$). In addition to the star's declination, we need the position of the observer. The counterpart of `SkyCoord` for celestial coordinates is `EarthLocation` (also imported in line 1), which allows us to set the geographical latitude and longitude of a location on Earth:

```
7   import astropy.units as u
8
9   # geographical position of the observer
10  obs = EarthLocation(lat=53*u.deg+28*u.arcmin+49*u.arcsec,
11                      lon=10*u.deg+14*u.arcmin+23*u.arcsec)
12
13  # get latitude
14  phi = obs.lat
```

In the expressions above, `u.deg` is equivalent to 1°, `u.arcmin` is 1', and `u.arcsec` is 1''. The `units` module from `astropy` implements rules to carry out computations in physical and astronomical unit systems. You will learn more about AstroPy units in Sect. 3.1.1.

Next we compute h with the help of trigonometric functions from the `math` module:

```
15  import math
16
17  h = math.acos(-math.tan(delta.radian) * math.tan(phi.radian))
```

It is necessary to convert the angles δ and ϕ to radians before applying the tangent function `math.tan()`. With Astropy coordinates, all we need to do is to use the `radian` attribute of an angle.[8] To obtain T in hours, it is important to keep in mind

---

[7]ICRS means International Celestial Reference System.

[8]You could also work with the `SkyCoord` object defined in line 3 and use `betelgeuse.dec.radian` as argument of `math.tan()`. `betelgeuse.dec` is an `Angle` object. So there are objects within objects. However, the function `math.radians()` for converting from degrees to radians is not compatible with `Angle` objects.

that an angle of 360° (one full rotation of Earth) corresponds to a sidereal day, which is about 4 min shorter than a solar day. As explained in [3, Sect. 2.13], this is a consequence of the orbital motion of Earth. The conversion is made easy by the `units` module (see also Exercise 2.5 for a poor man's calculation):

```
18  T = (math.degrees(2*h)/360)*u.sday
19  print("T = {:.2f}".format(T.to(u.h)))
```

First `T` is defined in sidereal days (`u.sday`), which is equivalent to 24^h or 360°. Then we convert to solar hours (`u.h`) by applying the method `to()` in line 19. The result is

```
    T = 13.31 h
```

If it were not for the Sun, Betelgeuse could be seen 13 h at the Observatory. Of course, the star will be visible only during the overlap between this period and the night, which depends on the date. We will return to this question in the following section.

The diurnal arc of the Sun plays a central role in our daily life, as it determines the period of daylight. In Sect. 2.1.1, we introduced an approximation for the declination $\delta_\odot$ of the Sun. By substituting the expression (2.1) for $\delta_\odot$ into Eq. (2.2), we can compute how the day length varies over the year. First we need to compute $\delta_\odot$ for N ranging from 0 to 364. Using NumPy, this is very easy. In the following example, the expression `np.arange(365)` fills an array with the sequence of integers starting from 0 up to the largest number *smaller than* 365, which is 364.[9] Apart from that, the code works analogous to the NumPy-based computation for equinoxes and solstices in Sect. 2.1.1. Since a new task begins here, the line numbering is reset to 1, although we make use of previous assignments (an example is the latitude `phi`). In other words, you would need to add pieces from above to make the following code work as an autonomous program (you might want to try this).

```
1  import numpy as np
2
3  N = np.arange(365) # array with elements 0,1,2,...,364
4  omega = 2*math.pi/365.24 # Earth's angular velocity in rad/day
5  ecl = math.radians(23.44) # obliquity of the ecliptic
6
7  # calculate declination of the Sun for all days of the year
8  delta = -np.arcsin(math.sin(ecl) * np.cos(omega*(N+10)))
```

Now we can compute the day length T for all values in the array `delta` using functions from the `numpy` module (compare to the code example for Betelgeuse and check which changes have been made):

---

[9]Just like `range()` in `for` loops, the general form is `np.arange(start, stop, step)`, where a start value different from 0 and a step other than 1 can be specified as optional arguments.

```
9   # calculate day length in solar hours
10  h = np.arccos(-np.tan(delta) * math.tan(phi.radian))
11  T = (np.degrees(2*h)/360) * u.sday.to(u.h)
```

Here, `phi` is still the latitude of Hamburg Observatory defined above. Of course, `T` is now an array with 365 elements. Since we want the day length in solar hours, we multiply right away with `u.sday.to(u.h)`, which is the length of a sidereal day in solar hours.

When dealing with large data sets (i.e. more than a few values), it is most of the time preferable to extract some statistics or to display the data in graphical form. To show the annual variation of the day length, producing a plot of T versus N is the obvious thing to do. The Python library `matplotlib` (see matplotlib.org) provides a module called `pyplot` for plotting data in arrays:

```
12  import matplotlib.pyplot as plt
13  %matplotlib inline
14
15  plt.plot(N, T)
16  plt.xlabel("Day")
17  plt.ylabel("Day length [hr]")
18  plt.savefig("day_length.pdf")
```

The function `plt.plot()`, where `plt` is a commonly used alias for `matplotlib.pyplot`, produces a plot showing data points (x, y) given by the arrays `N` (x axis) and `T` (y axis). By default, the points are joined by lines to produce a continuous graph. Axes labels are added in lines 16 and 17. The plot is then saved to a file called `day_length.pdf` in PDF format. The location of the file depends on the directory in which you started your Python session. You can specify full path if you want to store the plot somewhere else). You can also use other graphics formats, such as PNG or JPG, by specifying the corresponding extension in the filename (e.g. `day_length.png`).[10] In line 13, you can see a so-called magic command (indicated by `%` at the beginning of the line). It enables inline viewing of plots in IPython and Jupyter notebooks.[11]

The graph can be seen as solid line in Fig. 2.4 (the dot-dashed line will be added below). As expected, the day length is short in January, reaches a maximum at the first solstice ($N = 171$) and then decreases until the second solstice is reached at $N = 355$. The minimal and maximal day length can be inferred with the help of `min()` and `max()` methods for NumPy arrays[12]:

```
19  print("Minimum day length = {:5.2f} h".format(T.min()))
20  print("Maximum day length = {:5.2f} h".format(T.max()))
```

---

[10]PDF has the advantage of containing the plot as a vector graphics: The number of pixels is not fixed and, as a result, lines and fonts appear smooth even when viewed at higher graphical resolution.

[11]Depending on your Python environment, it might be necessary to add the line `plt.show()` at the bottom.

[12]Functions such as `np.arange()` or `np.tan()` are called with an array as argument, whereas `min()` and `max()` are methods called on a specific object (here, the array `T`).

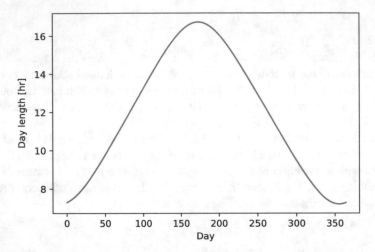

Fig. 2.4 Annual variation of the day length in Hamburg following from an approximate Eq. (2.1) for the declination of the Sun

We get

```
Minimum day length =   7.20 h
Maximum day length = 16.73 h
```

The difference between minimal and maximal day length increases with latitude. Beyond the polar circles ($\phi = \pm 66°\ 33'$), the day length varies between 0 and 24 h. An example is Longyearbyen located at $\phi = +78°\ 13'$ on the Norwegian island Spitsbergen in the far North. As we only need to know ϕ within an arc minute (accuracy is limited by the approximate declination anyway), a full specification of the geographical position using EarthLocation would rather overdo it. Instead we simply use $1' = (1/60)°$ and immediately convert into radians:

```
21  phi = math.radians(78+13/60) # latitude of Longyearbyen
22
23  h = np.arccos(-np.tan(delta)*math.tan(phi))
24  T = (np.degrees(2*h)/360)  * u.sday.to(u.h)
```

When you execute this code, you will get a

```
RuntimeWarning: invalid value encountered in arccos
```

It turns out that Python is able to handle this, but we should nevertheless try to understand what is wrong here. Remember that the range of the declination of the Sun is $-\epsilon_0 \leq \delta_\odot \leq \epsilon_0$. Considering Eq. (2.2), you can convince yourself that the right-hand side becomes smaller than -1 or greater than 1 if $|\phi| \geq 90° - \epsilon_0$, i.e. if the location is in the polar regions. In this case, Eq. (2.2) has no solution because the arccosine is undefined if the absolute value of its argument is greater than unity. This corresponds to the polar night or polar day during which the Sun never rises or sets. This can be fixed by setting $\cos h_{\text{set}}$ equal to ± 1 whenever the right-hand side of Eq. (2.2) falls outside of the interval $[-1, 1]$. Using Numpy, this is easily achieved by means of the `np.clip()` function:

```
24  tmp = np.clip(-np.tan(delta)*math.tan(phi), -1.0, 1.0)
25  h = np.arccos(tmp)
26  T = (np.degrees(2*h)/360) * u.sday.to(u.h)
```

Assigning the clipped right-hand side of Eq. (2.2) to a temporary array is only intended to highlight this step. We leave it as a little exercise for you to combine lines 24 and 25 into a single line of code and to produce and view the resulting graph for the day length. As expected, you will find that people on Longyearbyen have to cope with a polar night (all day dark) during winter, and a polar day lasting 24 h in summer.

Having computed the day length for Hamburg and Longyearbyen, it would be instructive to compare the graphs in a single plot. Obviously, we need the data for both locations at the same time, but in the example above the data for Hamburg were overwritten by the computation for Longyearbyen. A straightforward way of resolving this problem would be to use differently named array, but in Python there is a more elegant and convenient alternative. Apart from arrays in which items are referenced by index, data can be collected in a Python dictionary. Similar to a dictionary in the conventional sense, data items in a dictionary are referenced by keywords rather than a numerical index. The data items of the dictionary can be just anything, including arrays. Let us set up a dictionary associating locations with their latitudes:

```
27  phi = { 'Hamburg'        : obs.lat.radian,
28          'Longyearbyen'  : math.radians(78 + 13/60) }
```

In Python, a dictionary is defined by pairs of keywords and items enclosed in curly braces. Each keyword, which has to be a string (e.g. `'Hamburg'`), is separated by a colon from the corresponding item (the expression `obs.lat.radian` in the case of Hamburg) followed by a comma. How are individual items accessed? For example,

```
29  print(phi['Hamburg'])
```

prints the the latitude of Hamburg in radians. The syntax is similar to accessing array elements, except for the key word instead of an index. Adding new items to a dictionary is very easy. For example, the following assignments add New York and Bangkok to our dictionary:

```
30  phi['New York'] = math.radians(40 + 43/60)
31  phi['Bangkok']  = math.radians(13 + 45/60)
```

Indeed, the dictionary now has four items, which can be checked by printing `len(phi)`.

The following code prints all items, computes the day length for each location, and combines them in a single plot:

```
32 for key in phi:
33     print(key + ": {:.2f} deg".format(math.degrees(phi[key])))
34
35     h = np.arccos(np.clip(-np.tan(delta)*math.tan(phi[key]),
36                           -1.0, 1.0))
37     T = (np.degrees(2*h)/360) * u.sday.to(u.h)
38
39     plt.plot(N, T, label=key)
40
41 plt.xlabel("Day")
42 plt.xlim(0,364)
43 plt.ylabel("Day length [hr]")
44 plt.ylim(0,24)
45 plt.legend(loc='upper right')
46 plt.savefig("daylength.pdf")
```

The graphical output for the places

```
Hamburg: 53.48 deg
Longyearbyen: 78.22 deg
New York: 40.72 deg
Bangkok: 13.75 deg
```

can be seen in Fig. 2.5. As expected, the day length in Longyearbyen varies between 0 and 24 h, while people in Bangkok, which is in the tropics, experience a day length around 12 h over the whole year. The day length in the temperate zones increases by several hours from winter to summer.

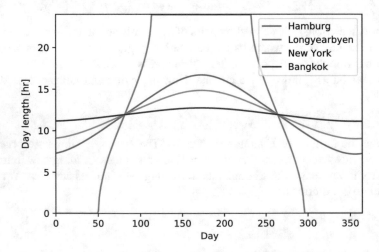

Fig. 2.5 Annual variation of the day length at different places in the world

To understand how this output is produced, let us go through the code step by step:

1. The code begins with a `for` loop iterating the dictionary `phi`. At first glance, this looks exactly like a loop through an array (see Sect. 2.1.1). However, there is an important difference. When iterating an array, the loop variable runs through the elements of the array. In contrast, the loop variable `key` runs through keywords, not the dictionary items themselves. Since the keyword is the analogue of the index, this loop resembles more an enumeration. The items are referenced by `phi[key]` in the loop body.
2. In line 33, the + operator concatenates two strings. The first string is a keyword, i.e. the name of a location, the second string is a formatted latitude.
3. Lines 35 to 37 correspond to lines 24–26 with the noticeable difference that `phi[key]` instead of `phi` refers to a particular latitude.
4. The last statement in the loop body (line 39) adds the graph for the current array `T`, which changes with each iteration, to the plot. This means that `plt.plot()` called inside a loop does not produce several plots, but accumulates graphs within a single plot. The optional argument `label=key` sets the graph's label to the keyword. The labels are used in the legend of the plot (see below).
5. Once the loop is finished, the plot is configured by the statements in lines 41–45: Axes are labeled, their range is limited by `plt.xlim()` and `plt.ylim()`, and labels for the different graphs are shown in a legend in the upper right corner of the plot.
6. Finally, the plot is saved to a PDF file. This will overwrite the previously saved file, unless you choose a different file name.

If you are unsure about any of the steps explained above, make changes to the code and see for yourself what happens and whether it meets your expectations.

2.1.3 Observation of Celestial Objects

While it is important to be able to specify the positions of celestial objects in a coordinate system that is independent of time and the observer's location, at the end of the day you want to know where on the sky you can find the object. To that end, the so-called horizontal coordinate system is used, which is based on an imaginary plane that is oriented tangential to the surface of Earth at the location of the observer. The angular position measured in normal direction from the horizon is the altitude a, and the angular separation from some reference direction (usually, the North direction) parallel to the horizon is the azimuth of the object.[13] This is why the horizontal system also goes under the name of alt-azimuth system. The module `astropy.coordinates` offers a powerful framework for working with celestial

---

[13] The azimuth measured from North corresponds to the compass direction.

coordinates (for more information, see docs.astropy.org/en/stable/coordinates). Similar to `SkyCoord` for coordinates in the equatorial system, the alt-azimuth system is represented by the class `AltAz`. In this section, we will utilize `AltAz` to infer the time of observability of the star Betelgeuse. In the course of doing so, we will touch upon some advanced aspects of Python. If you find it too complicated at this point, you can skip over this section and return later if you like.

To begin with, we define once more the position of the observer (see also Sect. 2.1.2). You are invited to adjust all settings to your own location and time.

```
1  import astropy.units as u
2  from astropy.coordinates import \
3      SkyCoord, EarthLocation, AltAz, get_sun
4
5  # geographical position of the observer
6  obs = EarthLocation(lat=53*u.deg+28*u.arcmin+49*u.arcsec,
7                      lon=10*u.deg+14*u.arcmin+23*u.arcsec)
8
9  # get latitude
10 phi = obs.lat
```

Astropy also has a module to define time in different formats.[14] By default, `Time()` sets the coordinated universal time (UTC).[15] To define, for example, noon in the CEST time zone (two hours ahead of UTC) on July 31st, 2020, we shift the time by an offset of two hours:

```
11 from astropy.time import Time
12
13 utc_shift = 2*u.hour   # CEST time zone (+2h)
14
15 noon_cest = Time("2020-07-31 12:00:00") - utc_shift
```

Printing `noon_cest` shows the UTC time corresponding to 12am CEST. The shift to UTC is necessary because `AltAz` expects UTC time.

With time and location, we can define our horizontal coordinate system. We want to follow the star's position over a full 24 h period and determine the time window for observation during night. Since altitude and azimuth are time dependent, we need to create a sequence of frames covering a whole day (frame is synonymous to coordinate system):

```
16 import numpy as np
17
18 # time array covering next 24 hours in steps of 5 min
19 elapsed = np.arange(0, 24*60, 5)*u.min
20 time = noon_cest + elapsed
21
```

---

[14]See docs.astropy.org/en/stable/time.

[15]The coordinated universal time (UTC) is an international standard for time. It is close (within a second) to the universal time that is related to Earth's rotation relative to distant celestial objects.

```
22   # sequence of horizontal frames
23   frame_local_24h = AltAz(obstime=time, location=obs)
```

In lines 19–20, a time sequence is created, starting from noon (CEST) in steps of five minutes covering an interval of 24 h. Let us look at this in more detail. First `np.arange()` creates an array of numbers $(0, 5, 10, \ldots)$. Calling

```
type(np.arange(0, 24*60, 5))
```

informs us that the type of the object returned by `np.arange()` is

```
numpy.ndarray
```

What is the type of an object? It is just the class to which the object belongs. Do not confuse this with the data type of the array (i.e. the type of its elements; see Sect. 2.1.1). As we shall see, the initially created NumPy array undergoes quite a metamorphosis. To define the time elapsed since noon in minutes, the array is multiplied with `u.min`. While multiplication with a number does not affect the type,

```
type(elapsed)
```

returns

```
astropy.units.quantity.Quantity
```

As mentioned in Sect. 1.4, quantities with units are instances of the `Quantity` class, which is defined in the submodule `astropy.units.quantity`. On the other hand, when printing `elapsed`, you will probably conclude that it pretty much looks like an array. Can we confirm this? The function **isinstance()** tells you if something belongs to a certain class (this is what instance means). While an object is of course an instance of the class defining its type,[16] it can also belong to other classes. Indeed,

```
isinstance(elapsed, np.ndarray)
```

yields `True`, i.e. `elapsed` is an array of sorts. Astropy supports arrays with units, which are a subclass of NumPy arrays. A subclass can extend attributes and methods of an existing base class, here `numpy.ndarray`. This is known as inheritance.

Now you might guess that `time` is also an array, created by adding a fixed value to the elements of `elapsed` (see line 20). But no,

```
isinstance(time, np.ndarray)
```

returns

---

[16] You can check this with **isinstance**(elapsed, u.quantity.Quantity). It is important to correctly reference the class in your namespace. Since we imported `astropy.units` using the alias u, we need to use the name `u.quantity.Quantity`. However, the function **type()** always returns the class in the generic module namespace.

```
False
```

Without going into too much detail, `time` has some array-like features, but it is not derived from NumPy. And it has its own way of treating units (basically, it replaces the notion of units by time formats). In general, it cannot be used in place of an array, such as `elapsed`. The reason we need it is to create frames by means of `AltAz()` (see line 23).

After this little detour to classes and inheritance, we retrieve Betelgeuse's coordinates as shown in Sect. 2.1.2 and then transform it to the observer's horizontal frame for the times listed in `time`.

```
28  # star we want to observe
29  betelgeuse = SkyCoord.from_name('Betelgeuse')
30
31  betelgeuse_local = betelgeuse.transform_to(frame_local_24h)
```

The method `transform_to()` turns the declination and right ascension of the star into altitudes and azimuths for the time sequence of frames defined in line 23 above. Modern telescopes are controlled by software, making it relatively easy for the observer to direct a telescope to a specific celestial object. Basically, the software uses something like Astropy's `transform_to()` and moves the telescope accordingly. Historical instruments, such as the 1 m reflector shown in Fig. 2.6, are a different matter. They are operated manually and the observer needs to know the position of an object on the sky.

To determine the phases of daylight, we need to determine the Sun's position in the same frames. Since the position of the Sun changes not only in the horizontal, but also in the equatorial coordinate system, Astropy offers a special function for the Sun:

```
32  # time-dependent coordinates of the Sun in equatorial system
33  sun = get_sun(time)
34
35  sun_local = sun.transform_to(frame_local_24h)
```

This completes the preparation of the data.

Now let us plot the altitude of Betelgeuse and the Sun for the chosen time interval and location:

```
36  import matplotlib.pyplot as plt
37  %matplotlib inline
38
39  elapsed_night = elapsed[np.where(sun_local.alt < 0)]
40  betelgeuse_night = \
41      betelgeuse_local.alt[np.where(sun_local.alt < 0)]
42
43  plt.plot(elapsed.to(u.h), sun_local.alt,
44          color='orange', label='Sun')
45  plt.plot(elapsed.to(u.h), betelgeuse_local.alt,
46          color='red', linestyle=':',
```

Fig. 2.6 Hamburg Observatory's 1 m reflector. When the instrument was commissioned in 1911, it was the fourth largest reflector worldwide. In the first half of the 20th century, the astronomer Walter Baade used the telescope to observe a great number of star clusters, gas nebulae, and galaxies. For further information, see www.physik.uni-hamburg.de/en/hs/outreach/historical/instruments.html (Image credit: Markus Tiemann, Martiem Fotografie)

```
47              label='Betelgeuse (daylight)')
48  plt.plot(elapsed_night.to(u.h), betelgeuse_night,
49              color='red', label='Betelgeuse (night)')
50
51  plt.xlabel('Time from noon [h]')
52  plt.xlim(0, 24)
53  plt.xticks(np.arange(13)*2)
54  plt.ylim(0, 60)
55  plt.ylabel('Altitude [deg]')
56  plt.legend(loc='upper center')
57  plt.savefig("Betelgeuse_obs_window.pdf")
```

Let us begin with the plot statement in lines 43–44. We plot the altitude of the Sun (`sun_local.alt`) as a function of the time elapsed since noon in hours. Rather than using default colors, we set the line color to orange (`color='orange'`). The graph for the altitude of Betelgeuse has two components. We use a dotted line (`linestyle=':'`) in red to plot the altitude over the full 24 hour interval (lines 45–47). Since the star is only observable during night, the dotted line is overplotted by a solid line for those times where the altitude of the Sun is negative, i.e. below the horizon—that is basically the definition of night. How do we do that? You find the answer in lines 39–41, where the expression

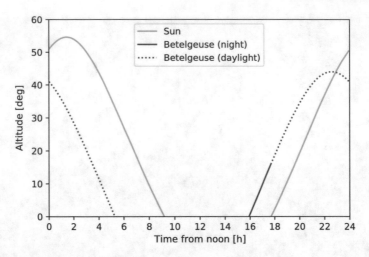

Fig. 2.7 Altitude of the Sun and the star Betelgeuse as seen from Hamburg Observatory on July 31st, 2020

```
np.where(sun_local.alt < 0)
```

appears like an index in brackets. The function `np.where()` identifies those elements of `sun_local.alt` which are less than zero and returns their indices (remember that array elements are identified by their indices). These indices in turn can be applied to select the corresponding elements in `elapsed` and `betelgeuse_local.alt`, producing *masked* arrays. This is equivalent to looping through the arrays, where any element `elapsed[i]` and `betelgeuse_local.alt[i]` for which `sun_local.alt[i]` is *not* smaller than zero (Sun above the horizon) is removed. In Exercise 2.4, you will learn how `np.where()` can be utilized to choose between elements of two arrays depending on a condition. This can be regarded as the array version of branching.

The result can be seen in Fig. 2.7. Since objects with negative altitude are invisible for the observer, only positive values are shown (see line 54). Moreover, the tick labels for time axis are explicitly set with `plt.xticks()` in line 53. Night is roughly between 9 pm and 6 am (18 h counted from noon on the previous day).[17] Betelgeuse rises around 4 o'clock in the morning and there only two hours left before sunrise. This is the interval for which the solid red line plotted in lines 48–49 is visible in the plot. Actually, the observation window is even narrower. Dawn is rather long at a latitude of 53°. The period of complete darkness, the so-called astronomical night, is defined by $\delta_\odot \leq -18°$. Although Betelgeuse is a very bright star, Hamburg is definitely not the optimal place to observe the star in summer (see Exercise 2.7).

---

[17] You might be surprised that the Sun does not reach its highest altitude at noon. One obvious reason is that summer time shifts noon (12 pm) by one hour. Moreover, depending on the longitude of the observer, time zones are not perfectly aligned with the Sun's culmination.

Exercises

2.1 Compute the Sun's declination for the equinoxes and solstices using only trigonometric functions from the `math` module in an explicit `for` loop. Print the results and check if they agree with the values computed with NumPy in this section. This exercise will help you understand what is behind an implicit loop.

2.2 The day count N in Eq. (2.1) can be calculated for a given date with the help of the module `datetime`. For example, the day of the vernal equinox in the year 2020 is given by

```
vernal_equinox = datetime.date(2020, 3, 20) - \
                 datetime.date(2020, 1, 1)
```

Then `vernal_equinox.days` evaluates to 79. Define the array N (equinoxes and solstices) using `datetime`.

2.3 A more accurate formula for the declination of the Sun takes the eccentricity $e = 0.0167$ of Earth's orbit into account[18]:

$$\delta_\odot = -\arcsin\left[\sin(\epsilon_0)\cos\left(\frac{360°}{365.24}(N+10) + e\frac{360°}{\pi}\sin\left[\frac{360°}{365.24}(N-2)\right]\right)\right]$$

Compute the declination assuming a circular orbit (Eq. 2.1), the declination resulting from the above formula, the difference between these values, and the relative deviation of the circular approximation in % for equinoxes and solstices and list your results in a table. Make sure that an adequate number of digits is displayed to compare the formulas.

2.4 The highest altitude a_{max} (also know as upper culmination) of a star measured from the horizontal plane[19] of an observer on Earth is given by

$$a_{max} = \begin{cases} 90° - \phi + \delta & \text{if } \phi \geq \delta, \\ 90° + \phi - \delta & \text{if } \phi \leq \delta, \end{cases} \tag{2.3}$$

where ϕ is the latitude of the observer and δ the declination of the star. Calculate a_{max} at your current location for the following stars: Polaris ($\delta = +89°\ 15'\ 51''$), Betelgeuse ($\delta = +07°\ 24'\ 25''$), Rigel ($\delta = -08°\ 12'\ 15''$), and Sirius A ($\delta = -16°\ 42'\ 58''$). To distinguish the two cases in Eq. (2.3), use the `where()` function from `numpy`. For example, the expression

```
np.where(phi <= delta, phi-delta, phi+delta)
```

---

[18] See en.wikipedia.org/wiki/Position_of_the_Sun#Calculations.

[19] The horizontal plane is tangential to Earth's surface at the location of the observer, assuming that the Earth is a perfect sphere.

compares `phi` and `delta` element by element and returns an array with elements of `phi-delta` where the condition `phi <= delta` is true and elements of `phi+delta` where it is false. Print the results with the appropriate number of significant digits together with the declinations.

2.5 A sidereal day is about 3 min 56 s shorter than a solar day (24 h). Show that this implies $1^h \approx 0.9973$ h. How would you need to modify the definition of `T` in Sect. 2.1.2 to make use of this factor without utilizing Astropy units?

2.6 Compute and plot the annual variation of the day length at your geographical location using both the formula with eccentricity correction from Exercise 2.3 and `get_sun()` from `SkyCoord`. How large is the deviation? How does your result compare with other places shown in Fig. 2.5?

2.7 Determine the observation window for Betelgeuse at New Year's Eve. Begin with the location of Hamburg Observatory (see Sect. 2.1.3). How many hours is the star observable during astronomical night, i.e. when the Sun is at least 18° below the horizon? Change to your location and compute the altitudes of Polaris, Betelgeuse, and Sirius A for the upcoming night. Produce plots similar to Fig. 2.4. Provided the sky is clear, which stars would you be able to see?

2.2 Kepler's Laws of Planetary Motion

The fundamental laws of planetary motion were first formulated by the astronomer Johannes Kepler on the basis of empirical data in the early 17th century. Kepler used the most accurate measurements of planetary positions available at that time. Particularly for the planet Mars, he noticed that its apparent motion on the sky can be explained by assuming that the planet follows an elliptical orbit around the Sun. This was quite a revolutionary proposition, as the motion of planets was considered to be circular (more precisely, a combination of circular motions in the Ptolemaic system), with Earth residing at the centre of the Universe. Although the differences were rather minute, Kepler concluded from his analysis that the planets move on elliptical orbits, where the Sun is located in one focal point (first law) and, in modern language, the area swept by the radial vector from the Sun to a planet is proportional to the elapsed time (second law). Later he found a relation between the semi-major axes (the line segment from the center through the focus to the perimeter of an ellipse) and the orbital periods, which today is known as Kepler's third law. Isaac Newton found a theoretical explanation for Kepler's laws of planetary motion by applying his universal law of gravity and angular momentum conservation (see [4, Chap. 2] for a detailed derivation and discussion). We will return to the first and second law in Sect. 4.3, when numerical integration is applied to solve the equations of motion for the two-body problem (Sun and planet).

The general formulation for the period P of a planet moving on an elliptical orbit around a star reads

$$P^2 = \frac{4\pi^2}{G(M+m)}\, a^3, \tag{2.4}$$

where a is the semi-major axis of the orbit, G the gravitational constant, M the mass of the star, and m the mass of the planet. In Sect. 1.2, we used an approximate expression (1.2), which is applicable if the planet's mass is negligible, i.e. $m \ll M$, and the orbit is circular (in this case, a is equal to the radius r).

The following program computes the orbital period given by Eq. (2.4) for the eight planets of the solar system. In fact, this is also an approximation because the mutual influence of planets is neglected. In other words, Kepler's third law assumes a two-body system (the Sun and a single planet). It is more accurate than the test-mass approximation ($m = 0$) though. The code compares the periods resulting from the two-body and test-mass formulas.

```
1  import math
2  import numpy as np
3  from scipy.constants import year,hour,au,G
4  from astropy.constants import M_sun
5
6  M = M_sun.value # mass of the Sun in kg
7
8  # orbital parameters of planets
9  # see https://nssdc.gsfc.nasa.gov/planetary/factsheet/
10 # mass in kg
11 m = 1e24 * np.array([0.33011, 4.8675, 5.9723, 0.64171,
12                      1898.19, 568.34, 86.813, 102.413])
13 # semi-major axis in m
14 a = 1e9 * np.array([57.9, 108.21, 149.60, 227.92,
15                     778.57, 1433.53, 2872.46, 4495.06])
16
17 # use Kepler's third law to calculate period in s
18 T_test_mass = 2*math.pi * (G*M)**(-1/2) * a**(3/2)
19 T_two_body = 2*math.pi * (G*(M + m))**(-1/2) * a**(3/2)
20
21 print("T [yr]   dev [hr] dev rel.")
22 for val1,val2 in zip(T_test_mass,T_two_body):
23     dev = val1 - val2
24     if dev > hour:
25         line = "{0:6.2f}   {1:<7.1f}   {2:.1e}"
26     else:
27         line = "{0:6.2f}   {1:7.4f}   {2:.1e}"
28     print(line.format(val2/year, dev/hour, dev/val1))
```

We use `scipy.constants` for the gravitational constant G and the definition of the mass of the Sun from `astropy.constants` (see Sect. 1.4). The masses of the planets and the semi-major axis of their orbits around the Sun are taken from

NASA's website (see comment). Calculations in the program above are based on the SI unit system. To express the results in years and astronomical units, we use unit conversion factors from `scipy.constants`. This factors are just numbers. For example, `year` is one year in seconds and `au` is one astronomical unit in meters. In Sect. 3.1, you will learn how to use Astropy's unit module to carry out unit conversions with the help of methods (a first taste was given in the previous section).

The orbital periods defined by Eq. (2.4) and the test-mass approximation, where the planet mass is neglected, are computed in lines 18 and 19, respectively, using array operations. The results are printed in the `for` loop in lines 22–28 (see Sect. 2.1 for an explanation of the `zip()` function), where the first column lists the periods given by Eq. (2.4), the second column the deviation from the test-mass approximation in hr, and the third column the relative deviation:

```
T [yr]   dev [hr] dev rel.
  0.24    0.0002  8.3e-08
  0.62    0.0066  1.2e-06
  1.00    0.0132  1.5e-06
  1.88    0.0027  1.6e-07
 11.88   49.6     4.8e-04
 29.68   37.2     1.4e-04
 84.20   16.1     2.2e-05
164.82   37.2     2.6e-05
```

While Mercury revolves around the Sun four times a year, it takes Neptune well above a century to complete its orbit. To print the period in years, we need to divide the value computed in SI units (s) by the conversion factor `year` (see line 28). The deviation from the test-mass approximation is larger for the outer planets (Jupiter, Saturn, Uranus, and Neptune), as they are much more massive than the inner planets (Mercury, Venus, Earth, and Mars). Nevertheless, the relative error of the orbital period is small. To align the values in the second column at the decimal point, an `if-else` clause is used to define the format of a line depending on whether the absolute deviation is larger or smaller than one hour (`hour` is one hour in seconds). The column width is 7 digits in both cases, but the values for the outer, more massive planets are printed with only one digit after the decimal point and shifted to the left by inserting the character < in the format specifier in line 25 (experiment with different formats to understand how they work).

Power laws, such as $P \propto a^{2/3}$, appear as straight lines with slope equal to the exponent in a double-logarithmic diagram:

$$\log P = \frac{2}{3} \log a + \text{const.} \tag{2.5}$$

We can plot our results in log-log scaling by applying the function `loglog()` from the `pyplot` module:

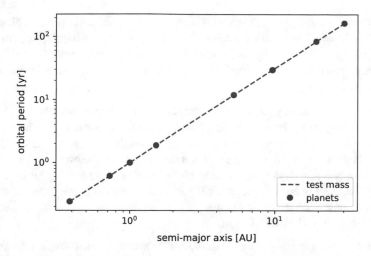

Fig. 2.8 Orbital periods of the planets in the solar system computed with Kepler's third law

```
29  import matplotlib.pyplot as plt
30  %matplotlib inline
31
32  plt.loglog(a/au, T_test_mass/year, 'blue', linestyle='--',\
33             label='test mass')
34  plt.loglog(a/au, T_two_body/year, 'ro', label='planets')
35  plt.legend(loc='lower right')
36  plt.xlabel("semi-major axis [AU]")
37  plt.ylabel("orbital period [yr]")
38  plt.savefig("kepler_third_law.pdf")
```

This graphical output is shown in Fig. 2.8. Plotting the period in hr versus the semi-major axis in AU is simply a matter of dividing the arrays by the `scipy` unit conversion factors `hr` (s to hr) and `au` (m to AU), respectively. The test-mass data, which obey the power law exactly, are displayed as blue dashed line, while the elements in `T_two_body` are plotted as red dots ('ro' is the short notation for red circle).[20] The keyword `color` (see Sect. 2.1.3) can be omitted if it is the *third* argument after the two data arrays that specifies the color of the graph. This type of argument is called positional argument, as opposed to an argument identified by a keyword such as `linestyle`. Different types of arguments will be covered in more detail in Sect. 3.1.2, when you learn how to define functions in Python.

Exercises

2.8 In addition to the eight planets in the solar system, there are several dwarf planets. Examples are Pluto ($a = 39.48$ AU), which was formerly considered as

---

[20]See https://matplotlib.org/stable/api/colors_api.html and
https://matplotlib.org/stable/api/markers_api.html for available colors and markers, respectively.

ninth planet,[21] Ceres ($a = 2.7675$ AU) in the asteroid belt, and the trans-Neptunian object Eris ($a = 67.781$ AU). Compute the corresponding orbital periods (the mass of the dwarf planets is negligible) and plot the results together with the orbital data of the planets using different markers and an additional label in the legend for the dwarf planets.

2.9 Since the discovery of the first exoplanets (extrasolar planets) in the mid-nineties, many more have been identified. An important class are the so-called hot Jupiters. These objects have high mass, but unlike the gas giants in the solar system, they are much closer to their parent star. Typically, the orbital period of a hot Jupiter is only a few days (compared to almost twelve years for Jupiter in the solar system). Table 2.1 lists a sample of hot Jupiters discovered with the transient method. The planet mass is often a lower bound, as it depends on the unknown inclination of the orbit relative to the line of sight.

(a) Compute the semi-major axis a in AU for these exoplanets using the planet mass m and the mass of the star M as parameters and plot the orbital period P versus a and plot your results.

(b) You can use the NumPy function `polyfit()` to determine a linear fit to the logarithmic data. Call `polyfit()` with the logarithm of P in days as first argument (x-data), the logarithm of a in AU as second argument (y-data), and degree of the polynomial function that is fitted to the data as third argument. Here, the degree is 1 for a linear function $y = c_1 x + c_0$. `polyfit()` returns the fit parameters c_0 and c_1. Compare to the logarithmic formulation (2.5) of Kepler's third law. Why is the slope not exactly reproduced? Display the orbital period following from the fit as a line in the plot from (a).

2.10 Imagine a manned spaceship is sent to Mars. Let us assume that the spaceship follows a Hohmann transfer trajectory (see Fig. 2.9). This is the most energy saving (i.e. requiring the least amount of propellant), albeit not the fastest option to reach Mars. The trajectory is formed by one half of an elliptical orbit around the Sun touching the orbits of Earth and Mars at its perihelion and aphelion, respectively.[22] In basic calculations, the planetary orbits are assumed to be circular, with radii $r_\oplus = 1$ AU (Earth) and $r_{\mars} = 1.524$ AU (Mars). In this case, the radial distances of the perihelion and aphelion are $r_p = r_\oplus$ and $r_p = r_{\mars}$, respectively. After launch, thrusters push the spaceship from Earth's orbit to the transfer trajectory. Compute the semi-major axis a_H of the elliptical transfer orbit and the velocity difference $\Delta v = v_p - v_\oplus$, where v_p is the velocity required to enter the orbit at perihelion and $v_\oplus$ is the orbital velocity of Earth. (Apply vis-viva Eq. (4.46) to compute v_p; see also

---

[21] Pluto was degraded to a dwarf planet in 2006, after new criteria for the definition of a planet were accepted by the International Astronomical Union. However, the debate whether Pluto is a planet or not has gained momentum again, particularly after the New Horizons space mission delivered fascinating images of this icy world at the outskirts of the solar system.

[22] The term orbit usually refers to a special kind of trajectory, namely the path of periodic motion around a gravitational center.

Table 2.1 List of orbital and stellar parameters for a sample of hot Jupiters from exoplanets.org. The name of the exoplanet is derived from a stellar catalogue or from the discovering instrument (the letter 'b' always indicates the first exoplanet in the system of the parent star), P is the orbital period in days, and m its mass in units of the Jupiter mass, $M_J = 1.898 \times 10^{27}$ kg. The mass M of the parent star is specified in units of the solar mass ($M_\odot = 1.988 \times 10^{30}$ kg)

Exoplanet	P/day	m/M_J	$M/M_\odot$
CoRoT-3 b[a]	4.257	22	1.37
Kepler-14 b	6.790	8.4	1.51
Kepler-412 b	1.721	0.94	1.17
HD 285507 b	6.088	0.92	0.73
WASP-10 b	3.093	3.19	0.79
WASP-88 b	4.954	0.56	1.45
WASP-114 b	1.549	1.77	1.29

[a]With a mass larger than $13 M_J$, CoRoT-3 b is a brown dwarf

Fig. 2.9 Schematic view of a Hohmann transfer trajectory (solid black) from the Earth orbit (blue) to the Mars orbit (red). The perihelion and aphelion velocity at the vertex points are $\mathbf{v}_p$ and $\mathbf{v}_a$, respectively. The positions of the planets at launch time are indicated by E and M, the final position by M'. The angle δ is the initial angular separation of Earth and Mars and $\Delta\varphi$ is the angle swept by Mars over the transfer time t_H

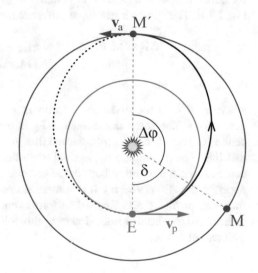

[4, Chap. 2] for important equations and relations.) How long is the transit time t_H to the orbit of Mars? Which condition for the angular separation δ of Earth and Mars (i.e. the angle between the position vectors $\mathbf{r}_\oplus$ and $\mathbf{r}_{\circlearrowleft}$) at launch time must be met for the spaceship to actually rendezvous with Mars at aphelion (assuming that Mars moves on a circular orbit)?

2.3 Tidal Forces

The orbital motion of two bodies is governed by Newton's law of gravitation for point masses. However, the dependence of the gravitational force on distance gives rise to tidal forces between different parts of an extended body, such as a planet, in the gravitational field of another body. Let us consider a small test mass m in the gravitational field $\mathbf{g} = (GM/r^2)\mathbf{e}_r$ of a distant body of mass M ($\mathbf{e}_r = \mathbf{r}/r$ is the unit vector in radial direction from the center of that body). The difference between the gravitational forces exerted on the test mass at neighboring points $\mathbf{r}$ and $\mathbf{r} + d\mathbf{r}$ can be expressed by the field gradient:

$$dF = m\nabla \mathbf{g} \cdot d\mathbf{r} = -\frac{2GMm}{r^3}dr \ . \tag{2.6}$$

Now, if we think of different parts of a planet rather than a test mass placed at different positions, the tidal force is given by the force difference with respect to the center of mass.

Virtually everybody knows about the tides due to the gravity of the Moon exerted on Earth. Consider a point P at distance $R \leq R_E$ from the center C of Earth (see Fig. 2.10). The approximate force difference is given by

$$\begin{aligned} \Delta\mathbf{F} \equiv (\Delta F_x, \Delta F_y) &= \mathbf{F}_{PM} - \mathbf{F}_{CM} \\ &\simeq \frac{GMmR}{r^3}(2\cos\theta, -\sin\theta) \ , \end{aligned} \tag{2.7}$$

where $r \gg R$ is the distance of Earth from the Moon (see [4, Sect. 19.2] for a derivation). The force component ΔF_x is directed along the line connecting the centers of mass C and M (dot-dashed line in Fig. 2.10) and ΔF_y is perpendicular to that line. The magnitude of the tidal force increases with the distance from Earth's center. It has a maximum both in the direction toward the Moon ($\theta = 0$) and in the opposite direction ($\theta = \pi$). It is sometimes conceived as counter-intuitive that tidal forces are strongest and directed outwards and, thus, producing high tides both at the nearest and the farthest side. However, this follows from the fact that tidal forces are differential forces.

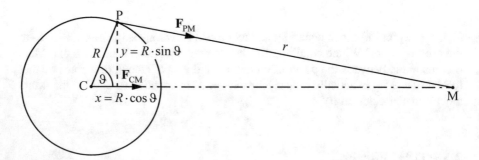

Fig. 2.10 Illustration of local gravitational forces exerted by the Moon (M) at the center of Earth (C) and at some point at the surface (P). The difference between the forces is the tidal force given by Eq. (2.7)

The following program computes the tidal force per unit mass, $\mathbf{a}_{\text{tidal}} = \Delta\mathbf{F}/m$, for a grid of points with equal spacing along the x- and y-axes within a circle of radius $R = R_{\text{E}}$:

```
1   import numpy as np
2   from scipy.constants import g,G
3   from astropy.constants import R_earth,M_earth
4
5   M = 0.07346e24 # mass of the moon in kg
6   r = 3.844e8 # semi-major axis of moon orbit in m
7
8   coeff = G*M/r**3
9   accel_scale = 2*coeff*R_earth.value
10  print("tidal acceleration = {:.2e} m/s^2 = {:.2e} g".\
11          format(accel_scale,accel_scale/g))
12
13  h = 15*M*R_earth.value**4/(8*M_earth.value*r**3)
14  print("size of tidal bulge = {:.2f} m".format(h))
15
16  # array of evenly spaced grid points along x- and y-axis
17  X = np.linspace(-1.1, 1.1, num=23, endpoint=True)
18  Y = np.linspace(-1.1, 1.1, num=23, endpoint=True)
19  print(X)
20
21  # create two-dimensional mesh grid scaled by Earth radius
22  R_x, R_y = np.meshgrid(R_earth.value*X, R_earth.value*Y)
23  print(R_x.shape)
24  print(R_x[11,21],R_y[11,21])
25
26  # radial distances of mesh points from (0,0)
27  R = np.sqrt(R_x*R_x + R_y*R_y)
28
29  # components of tidal acceleration field within Earth radius
30  accel_x = np.ma.masked_where(R > R_earth.value, 2*coeff*R_x)
31  accel_y = np.ma.masked_where(R > R_earth.value, -coeff*R_y)
```

The magnitude of $\mathbf{a}_{\text{tidal}}$ for $R = R_{\text{E}}$ along the axis connecting the centers of Earth and Moon (i.e. for $\theta = 0$) defines the scale $2GMR_{\text{E}}/r^3$ of the tidal acceleration. Its numerical value is calculated in lines 8 and 9 with constants from astropy.constants and data for the Moon[23]:

```
tidal acceleration = 1.10e-06 m/s^2 = 1.12e-07 g
```

Relative to the gravity of Earth, $g \approx 9.81$ m/s^2 (we use the constant g defined in scipy.constants), the tidal acceleration is very small. Otherwise tidal forces would have much more drastic effects on Earth. In line 13, we calculate the height of the tidal bulge of Earth caused by the Moon, using an approximate formula neglecting the rigidity of Earth [5]:

---

[23] nssdc.gsfc.nasa.gov/planetary/factsheet.

$$h = \frac{3M R_\mathrm{E}^4}{4 M_\mathrm{E} r^3}\zeta, \quad \text{where } \zeta \simeq 5/2. \tag{2.8}$$

The result is indeed comparable to the high tides in the oceans[24]:

```
size of tidal bulge = 0.67 m
```

The next step is the discretization of x and y coordinates by introducing grid points $x_n = n\Delta x$ and $y_n = n\Delta y$ along the coordinate axes. In lines 17 and 18, NumPy arrays of grid points are produced with np.linspace(). This function returns a given number of evenly spaced points in the interval specified by the first two arguments. We use dimensionless coordinates normalized by Earth's radius, so that we do not need to worry about actual distances. Since we want to cover a region somewhat larger than Earth, which has a diameter of 2.0 in the normalized coordinate system, we subdivide the interval $[-1.1, 1.1]$ into 23 points including both endpoints. This implies $\Delta x = \Delta y = 2.2/(23 - 1) = 0.1$, corresponding to a physical length of $0.1 R_\mathrm{E}$. The print statement in line 19 shows that the resulting elements of X:

```
[-1.1 -1.   -0.9 -0.8 -0.7 -0.6 -0.5 -0.4 -0.3 -0.2 -0.1  0.
  0.1  0.2  0.3  0.4  0.5  0.6  0.7  0.8  0.9  1.    1.1]
```

For a representation of the vector field $\mathbf{a}_\mathrm{tidal}(\mathbf{R})$, we need to to construct a two-dimensional mesh of position vectors $\mathbf{R} = (R_x, R_y) \equiv (x, y)$ in the xy plane. Owing to the rotation symmetry of the Earth-Moon system, we can disregard the z-component. A two-dimensional mesh grid of points can be constructed from all possible combinations of x_n and y_m with independent indices n and m. This is the purpose of the NumPy function meshgrid() called in line 22.   To obtain physical distances, the normalized coordinates are scaled with the radius of Earth (imported from astropy.constants). Since X and Y each have 23 elements, np.meshgrid() returns two-dimensional arrays defining the components R_x and R_y for a mesh grid consisting of $23 \times 23 = 529$ points. Here, multiple return values (in this case, arrays) are assigned to multiple variables (R_x and R_y). You can think of a two-dimensional array as a matrix, in this case with 23 rows and 23 columns. This can be seen by printing the shape of R_x (see line 23), which is

```
(23, 23)
```

To get a particular element, you need to specify *two* indices in brackets, the first being the row index and the second the column index. For example, the values of R_x[11,21] and R_y[11,21] printed in line 24 are

```
6378100.0 0.0
```

---

[24]Tides become much higher in coastal waters and the tides produced by the Sun amplify or partially compensate the lunar tides depending on the alignment with the Moon.

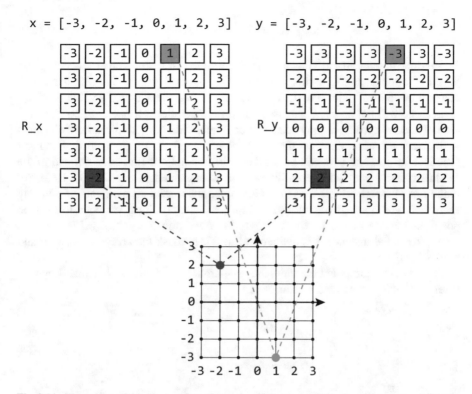

Fig. 2.11 Illustration of the construction of a mesh grid from one-dimensional arrays X and Y (top). The two-dimensional arrays R_x and R_y (middle) represent the coordinates of all points on the mesh (bottom). As examples, the coordinates of red and green dots are highlighted in the arrays R_x and R_y

i.e. the position is $\mathbf{R} = (R_{\mathrm{E}}, 0)$. To understand how this comes about, it might help to consider a simpler example with a smaller number of points and without scaling. Figure 2.11 shows how a mesh grid consisting of 7×7 points is constructed from one-dimensional arrays of size 7. The green dot, for example, has coordinates $(1, -3)$, which are given by R_x[0,4] and R_y[0,4] (see green squares in the two arrays in the middle of the figure). As you can see, the x-coordinate changes with the column index and the y-coordinate with the row index. You may wonder why the two dimensional arrays R_x and R_y are required. They are just redundant representations, are they not? After all, the rows of R_x are identical and so are the columns of R_y. However, the two-dimensional arrays allow us to immediately infer the components of the position vector (or some other vector) for *all* points on the mesh. In effect, this is a discrete representation of a vector field, where a vector is attached to each point in space (the field of position vectors is just a special case).

For a given position (R_x, R_y), the tidal acceleration $\mathbf{a}_{\mathrm{tidal}} = \Delta \mathbf{F}/m$ follows from Eq. (2.7), where $R\cos\theta = R_x$ and $R\sin\theta = R_y$. Thus,

$$\mathbf{a}_{\mathrm{tidal}}(\mathbf{R}) = \frac{GM}{r^3}(2R_x, -R_y) \ . \tag{2.9}$$

This is done in lines 30 and 31 using NumPy arithmetics. To constrain the acceleration field to a circular area of radius R_{E}, we define masked arrays, A mask flags each element of an array with 0 or 1. If an element is flagged with 1, its value becomes invalid and will not be used in subsequent array operations. Since masked elements are not removed from an array (they can be unmasked later), all unmasked elements with flag 0 are accessible via the same indices as in the original array. For most practical purposes, arrays can be masked with the help of functions defined in the module `numpy.ma` (an alternative method is explained in Sect. 2.1.3). Here, we make use of `masked_where()`. This functions masks all elements of an array for which which a logical condition evaluates to `True`. Since we want to exclude positions outside of the Earth, we apply the condition `R > R_earth.value`, where the radial distance `R` is defined in line 27, to mask the arrays `accel_x` and `accel_y`.

After having computed the data, the acceleration field can be visualized by representing vectors graphically as arrows:

```
32   import matplotlib.pyplot as plt
33   from matplotlib.patches import Circle
34   %matplotlib inline
35
36   fig, ax = plt.subplots(figsize=(6,6))
37   ax.set_aspect('equal')
38
39   # plot vector field
40   arrows = ax.quiver(X, Y, accel_x, accel_y, color='blue')
41   ax.quiverkey(arrows, X=0.1, Y=0.95, U=accel_scale,
42                label=r'$1.1\times 10^{-6}\;\mathrm{m/s}^2$',
43                labelpos='E')
44
45   # add a circle
46   circle = Circle((0, 0), 1, alpha=0.2, edgecolor=None)
47   ax.add_patch(circle)
48
49   ax.set_xlabel(r'$x/R_{\mathrm{E}}$', fontsize=12)
50   ax.set_ylabel(r'$y/R_{\mathrm{E}}$', fontsize=12)
51
52   plt.show()
53   plt.savefig("tidal_accel_earth.pdf")
```

A field of arrows can be produced with the method `quiver()`, which is called on an axes object in line 40. This object is created by invoking `plt.subplot()` (see line 36).[25] In line 37, the aspect ratio of the x and y-axes is set to unity to produce a quadratic plot. The call of `quiver()` requires the grid points `X` and `Y` along the coordinate axes and the meshgrid values `accel_x` and `accel_y` of the acceleration

---

[25]Usually the function `plt.subplots()` is applied to produce multiple plots in a single figure, see https://matplotlib.org/stable/api/_as_gen/matplotlib.pyplot.subplots.html and Sect. 4.2.

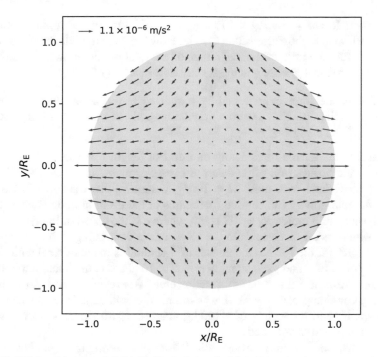

Fig. 2.12 Tidal acceleration field inside Earth due to the Moon's gravity

field as arguments. While X and Y determine the arrow positions (the meshgrid is constructed implicitly), their length and orientation is given by `accel_x` and `accel_y`. The arrow color can be specified by an optional argument, similar to the line color in `plt.plot()`. The `quiverkey()` method (line 41) displays an arrow labeled with the value corresponding to the length of the arrow at the coordinates specified in the argument list. This arrow is not part of the vector field we want to visualize; it is only meant to show the scale of the acceleration field. The interior of Earth is highlighted with the help of `Circle()` from `matplotlib.patches`. In our example, it produces a filled circle with unit radius (remember that coordinates in X and Y are normalized by Earth's radius) centered at position (0, 0). Experiment with the arguments of this function to see how they affect the appearance of the circle. To display the circle, `add_patch()` (line 47) has to be called on `ax`. This method inserts flat geometric objects into a plot (such objects are called patches in `matplotlib`). The resulting plot is shown in Fig. 2.12. It shows the typical pattern resulting in tidal bulges at opposite sides along the line between Earth and Moon. Since Earth is approximately a rigid body, bulges are induced only in the water of the oceans, giving rise to flood and low tide.

Exercises

2.11 Not only causes the Moon tidal forces on Earth, but also vice versa. Compare the tidal effect of Jupiter (mass and radius are defined in `astropy.constants` on its moon Io ($M = 8.9319 \times 10^{22}$ kg, $R = 1822$ km, mean orbital radius $r = 4.217 \times 10^5$ km) to the Earth-Moon system. How large are the tidal bulges of Io and the Moon? Plot the ratio of the magnitude of the tidal acceleration a_{tidal} defined by Eq. (2.9) to the local gravity g at the surface as a function of θ.[26]

2.12 The tensile force experienced by a cylindrical rod of length l and mass m directed in radial direction toward a gravitating body of mass M is found by integrating Eq. (2.7) using $l \ll r$.

(a) Estimate the tensile force acting on a rod of length $l = 1$ m at the surface of Earth, the surface of a white dwarf of one solar mass, and at the event horizon of a black hole with mass $M = 10 M_\odot$, assuming that the formula based on Newtonian gravity can be used (the radius of the event horizon is given by the Schwarzschild radius $R_S = 2GM/c^2$, where c is the speed of light).

(b) At which radial distance from the black whole is it going to be torn apart by tidal forces if the rod has a diameter of 5 cm and is made of steel with density $\rho = 7.8$ g cm^{-3} and yield strength $\sigma = 5 \times 10^8$ Pa (i.e. the maximum force the rod can resist is σ times the cross section area). Also estimate how close a human being could approach the black whole until it would experience a tensile force comparable to the weight of a 100 kg mass on Earth (imagine such a weight being attached to your body).

(c) Since any object falling toward a black hole and passing the event horizon will sooner or later experience extreme tidal forces, the radial stretching and compression in transversal directions would result in "spaghettification". Solid bodies, however, will be torn into ever smaller pieces due their limited deformability. Produce a plot similar to Fig. 2.12, showing the tidal acceleration field acting on the rod at the critical distance determined in (b). Use the function `Rectangle()` from `matplotlib.patches` to show the cross section of the rod along its axis.

---

[26]The surface gravity of a spherical body of mass M and radius R is given by $g = GM/r^2$.

Chapter 3
Functions and Numerical Methods

Abstract Topics such as black body radiation and Balmer lines set the stage for defining your own Python functions. You will learn about different ways of passing data to functions and returning results. Understanding how to implement numerical methods is an important goal of this chapter. For example, we introduce different algorithms for root finding and discuss common pitfalls encountered in numerics. Moreover, basic techniques of numerical integration and differentiation are covered. Generally, it is a good idea to collect functions in modules. As an example, a module for the computation of planetary ephemerides is presented.

3.1 Blackbody Radiation and Stellar Properties

Paradoxical as it may sound, the light emitted by a star is approximately described by the radiation of a black body.[1] This is to say that stars emit thermal radiation produced in their outer layers, which have temperatures between a few thousand and tens of thousands of K, depending on the mass and evolutionary stage of the star. The total energy emitted by a black body per unit time is related to its surface area and temperature. This relation is called the Stefan–Boltzmann law. The spectral distribution of the emitted radiation is given by Planck function. See Sect. 3.4 in [4] for a more detailed discussion of the physics of black bodies. In this section, Python functions are introduced to perform calculations based on the Stefan–Boltzmann law and the Planck spectrum and to discuss basic properties of stars.

---

[1] The term black body refers to its property of perfectly absorbing incident radiation at all wavelengths. At the same time, a black body is an ideal emitter of thermal radiation.

© Springer Nature Switzerland AG 2021   55
W. Schmidt and M. Völschow, *Numerical Python in Astronomy and Astrophysics*,
Undergraduate Lecture Notes in Physics,
https://doi.org/10.1007/978-3-030-70347-9_3

3.1.1 Stefan–Boltzmann Law

A star is characterised by its effective temperature T_{eff} and luminosity L (i.e. the total energy emitted as radiation per unit time). The effective temperature corresponds to the temperature of a black body radiating the same energy per unit surface area and unit time over all wavelengths as the star. This is expressed by the Stefan–Boltzmann law:

$$F = \sigma T_{\text{eff}}^4 , \tag{3.1}$$

where $\sigma = 5.670 \times 10^{-8}$ W m^{-2} K^{-4} is the Stefan–Boltzmann constant. The radiative flux F is the net energy radiated away per unit surface area and unit time. Integrating the radiative flux over the whole surface of a star of radius R, we obtain the luminosity

$$L = 4\pi R^2 \sigma T_{\text{eff}}^4 . \tag{3.2}$$

Suppose you want to compute the luminosity of a star of given size and effective temperature. You could do that by writing a few lines of Python code, just like the computation of the orbital velocity in Chap. 1. However, it is very common to write a piece of code in such a way that is re-usable and can perform a particular action or algorithm for different input data. This can be achieved by defining a Python function. We have already introduced many library functions, so you should be familiar with using functions by now. A new function can be defined with the keyword `def` followed by a name that identifies the function and a list of arguments in parentheses. Similar to loops, the function header ends with a colon and the indented block of code below the header comprises the body of the function. In the function body, the arguments are processed and usually, but not always, a result is returned at the end. As an example, consider the following definition of the function `luminosity()`.

```python
from math import pi
from scipy.constants import sigma # Stefan-Boltzmann constant

def luminosity(R, Teff):
    """
    computes luminosity of a star
    using the Stefan-Boltzmann law

    args: R - radius in m
          Teff - effective temperature in K

    returns: luminosity in W
    """
    A = 4*pi * R**2 # local variable for surface area
    return A * sigma * Teff**4
```

The function is explained by a comment enclosed in triple quotes, which is more convenient for longer comments and is used as input for `help()` to display information about functions, particularly if they are defined inside modules. The idea is that somebody who wants to apply the function, can type `help(luminosity)` to get instructions.[2] The variables R and Teff are called *formal* arguments because their values are not specified yet. The definition of the function applies to any values for which the arithmetic expressions in the function body can be evaluated.

A function is executed for particular data in a function call. As you know from previous examples, a function call can be placed on the right side of an assignment statement. For example, the function defined above is used in the following code to compute the luminosity of the Sun.

```
16  from astropy.constants import R_sun, L_sun
17
18  Teff_sun = 5778 # effective temperature of the Sun in K
19
20  print("Solar luminosity:")
21
22  # compute luminosity of the Sun
23  L_sun_sb = luminosity(R_sun.value, 5778)
24  print("\t{:.3e} W (Stefan-Boltzmann law)".format(L_sun_sb))
25
26  # solar luminosity from astropy
27  print("\t{:.3e} ({:s})".format(L_sun, L_sun.reference))
```

Here, the *actual* arguments R_sun.value (the value of the solar radius defined in astropy.constants) and Teff_sun (the effective temperature of the Sun defined in line 18) are passed to the function luminosity() and the code in the function body (lines 14 to 15) is executed with actual arguments in place of the formal arguments. After the expression following the keyword **return** in line 15 has been evaluated, the resulting value is returned by the function and assigned to L_sun_sb in line 23. For comparison, the observed luminosity from the astropy library is printed together with the luminosity resulting from the Stefan–Boltzmann law:

```
Solar luminosity:
3.844e+26 W (Stefan-Boltzmann law)
3.828e+26 W (IAU 2015 Resolution B 3)
```

The values agree within 1%. The reference for the observed value is printed in the above example with the help of the attribute reference of L_sun.

The variable A defined in the body of the function luminosity() is a *local* variable, which is not part of Python's global namespace. It belongs to the local

---

[2] Of course, for a single function, the user could just as well look at the code defining the function. But for a module containing many different functions, using `help()` is more convenient.

namespace of a function, which exists only while a function call is executed. If you try to print its value after a function call, you will encounter an error because its scope is limited to the function body:

```
28  print(A)
```

reports

```
    NameError: name 'A' is not defined
```

In contrast, variable names such as L_sun_sb or sigma in the above example are defined in the global namespace and can be used anywhere, including the body of a function. However, referencing global variables within a function should generally be avoided. It obscures the interface between data and function, makes the code difficult to read and is prone to programming errors. Data should be passed to a function through its arguments. A common exception are constant parameters from modules. For example, we import sigma from scipy.constants into the global namespace and then use it inside of the function luminosity() in line 13. To sum up[3]:

> Python functions have a local namespace. Normally, a function receives data from the global namespace through arguments and returns the results it produces explicitly.

Perhaps you noticed that the unit of the solar luminosity is printed for both variables, although the character W (for Watt) is missing in the output string in line 27. This can be understood by recalling that L_sun is an Astropy object which has a physical unit incorporated (see Sect. 2.1.2). Although the format specifier refers only to the numeric value, some magic (i.e. clever programming) built into **format**() and **print**() automatically detects and concatenates the unit to the formatted numerical value.

In the example above, we defined L_sun_sb as a simple float variable. By means of the units module we can assign dimensional quantities to variables in the likeness of Astropy constants. To obtain the same output as above without explicitly printing the unit W, we just need to make a few modifications:

---

[3]The preferred way of implementing Python functions we recommend in this book is in accordance with *call by value*. This is not built into the language. It is a choice that is up to the programmer. In fact, if a mutable object is passed to a function, any changes made inside the function via object methods will persist outside (call by object reference). In particular, this applies to array elements. However, this kind of manipulation can become quite confusing and is prone to errors unless you are an experienced Python programmer. Appendix A explains how to use calls by object reference in object-oriented programming.

```
1  from astropy.constants import R_sun, L_sun, sigma_sb
2  import astropy.units as unit
3
4  def luminosity(R, Teff):
5      """
6      function computes luminosity of star
7      using the Stefan-Boltzmann law with units
8
9      args: dimensinoal variables based on astropy.units
10              R - radius
11              Teff - effective temperature
12
13      returns: luminosity
14      """
15      A = 4*pi * R**2 # local variable for surface area
16      return sigma_sb * A * Teff**4
17
18  Teff_sun = 5778*unit.K
19
20  # compute luminosity from dimensional variables
21  L_sun_sb = luminosity(R_sun, Teff_sun)
22  print("\t{:.3e} (Stefan-Boltzmann law)".format(L_sun_sb))
```

First, `sigma_sb` from `astropy.constants` defines the Stefan–Boltzmann constant with units (simply print `sigma_sb`). Second, a physical unit is attached to the variable `Teff_sun` by multiplying the numerical value `5778` with `unit.K` in line 17. Then `luminosity()` is called with the full object `R_sun` rather than `R_sun.value` as actual argument. Arithmetic operators also work with dimensional quantities in place of pure floating point numbers. However, the flexibility has a downside: there is no safeguard against combining dimensional and dimensionless variables and you might end up with surprising results if you are not careful.

As introduced in Sect. 2.1.2, the method `to()` allows us to convert between units. For example, we can print the luminosity in units of erg/s without bothering about the conversion factor:

```
23  # convert from W to erg/s
24  print("\t{:.3e} (Stefan-Boltzmann law)".
25         format(L_sun_sb.to(unit.erg/unit.s)))
```

This statement prints

```
        3.844e+33 erg / s (Stefan-Boltzmann law)
```

It is even possible to combine different unit systems. Suppose the radius of the Sun is given in units of km rather than m:

```
26  # compute luminosity with solar radius in km
27  L_sun_sb = luminosity(6.957e5*unit.km, Teff_sun)
```

```
28    print("\t{:.3e} (Stefan-Boltzmann law)".
29          format(L_sun_sb.to(unit.W)))
```

It suffices to convert the result into units of W without modifying the function
`luminosity()` at all (also check what you get by printing `L_sun` directly):

```
      3.844e+26 W (Stefan-Boltzmann law)
```

Nevertheless, complications introduced by Astropy units might sometimes out-
weigh their utility. As in many examples throughout this book, you might find it easier
to standardize all input data and parameters of a program to a fixed unit system such
as SI units. In this case, dimensional quantities are implied, but all variables in the
code are of simple floating point type. Depending on your application, you need to
choose what suits you best.

Having defined the function `luminosity()`, we can calculate the luminosity of
any star with given radius and effective temperature.[4] We will perform this calculation
for a sample of well known stars, namely Bernard's Star, Sirius A and B, Arcturus,
and Betelgeuse. The straightforward way would be to define variables for radii and
temperatures and then call `luminosity()` with these variables as arguments. In
the following, this is accomplished by a more sophisticated implementation that
combines several Python concepts:

```
30    def stellar_parameters(*args):
31        '''
32        auxiliary function to create a dictionary
33        of stellar parameters in SI units
34
35        args: (radius, effective temperature)
36        '''
37        return { "R"    : args[0].to(unit.m),
38                 "Teff" : args[1].to(unit.K) }
39
40    # dictionary of some stars
41    stars = {
42        'Bernard\'s Star' :
43            stellar_parameters(0.196*R_sun, 3.13e3*unit.K),
44        'Sirius A' :
45            stellar_parameters(1.711*R_sun, 9.94e3*unit.K),
46        'Sirius B' :
47            stellar_parameters(5.8e3*unit.km, 2.48e4*unit.K),
48        'Arcturus' :
49            stellar_parameters(25.4*R_sun, 4.29e3*unit.K),
50        'Betelgeuse' :
51            stellar_parameters(6.4e8*unit.km, 3.59e3*unit.K)
```

---

[4]Usually, this works the other way around: the radius can be calculated if the luminosity and the
temperature of a star are known from observations; see Exercise 3.1.

```
52  }
53
54  print("Luminosities of stars (relative to solar luminosity):")
55  for name in stars:
56      stars[name]['L'] = \
57          luminosity(stars[name]['R'], stars[name]['Teff'])
58      print("\t{:15s} {:.1e} ({:.1e}) ".format(name, \
59              stars[name]['L'], stars[name]['L']/L_sun))
```

First turn your attention to the dictionary defined in lines 41–52 (dictionaries are introduced in Sect. 2.1.2). The keywords are the names of the stars. The items belonging to these keys are returned by the function `stellar_parameters()` defined at the beginning of the program. As you can see in lines 37–38, the function returns a dictionary, i.e. each item of the dictionary `stars` is in turn a dictionary. Such a data structure is called a nested dictionary. As with any dictionary, new items can be added to the subdictionaries. This is done when the luminosity is calculated for each star by iterating over the items of `stars` and adding the result as a new subitem with the key `'L'` (lines 56–57). Items and subitems in nested dictionaries are referenced by keys in concatenated brackets (this syntax differs from multidimensional NumPy arrays). For example, `stars['Sirius B']['R']` is the radius of Sirius B and `stars['Sirius B']['L']` its luminosity (print some items to see for yourself). The whole dictionary of stellar parameters for Sirius B is obtained with `stars['Sirius B']`.

Before we look at the results, let us take a closer look at the function `stellar_parameters()`. In contrast to the function `luminosity()`, which has an explicit list of named arguments, `stellar_parameters()` can receive an arbitrary number of actual arguments. The expression `args*` just serves as a dummy. Such arguments are known as variadic arguments and will be discussed in more detail in Sect. 4.1.2. The function `stellar_parameters()` merely collects values in a dictionary provided that the number of arguments in the function call matches the number of keys (no error checking is made here because it serves only as an auxiliary for constructing the `stars` dictionary). Individual variadic arguments are referenced by an index. The first argument (`args[0]`) defines the radius and the second one (`args[1]`) the effective temperature of the star. Both values are converted into SI units. This implies that the function expects dimensional values as arguments, which is indeed the case in subsequent calls of `stellar_parameters()`. Generally, using an explicit argument list is preferable for readability, but in some cases a flexible argument list can be convenient. In particular, we avoid duplicating names in keys and arguments and new parameters and can be added easily.

The results are listed in the following table (`name` is printed with the format specifier `15s`, meaning a string with 15 characters, to align the names of the stars).

```
Luminosities of stars (relative to solar luminosity):
    Bernard's Star  1.3e+24 W (3.3e-03)
    Sirius A        9.9e+27 W (2.6e+01)
```

```
Sirius B          9.1e+24 W (2.4e-02)
Arcturus          7.5e+28 W (2.0e+02)
Betelgeuse        4.8e+31 W (1.3e+05)
```

We print both the luminosities in W and the corresponding values of $L/L_\odot$ (identify the corresponding expressions in the code). Bernard's Star is a dim M-type star in the neighbourhood of the Sun. With a mass of only $0.14 M_\odot$, it is small and cool. In contrast, Sirius A is a main sequence star with about two solar masses and spectral type A0. Thus, it is much hotter than the Sun. Since the luminosity increases with the fourth power of the effective temperature, its luminosity is about 26 times the solar luminosity. The luminosity of the companion star Sirius B is very low, yet its effective temperature 2.5×10^4 K is much higher than the temperature of Sirius A. Astronomers were puzzled when this property of Sirius B was discovered by Walter Adams in 1915. A theoretical explanation was at hand about a decade later: Sirius B is highly compact star supported only by its electron degeneracy pressure. The size of such a white dwarf star is comparable to size of the Earth,[5] while its mass is about the mass of the Sun. Owing to its small surface area (see Eq. 3.2), Sirius B has only a few percent of the solar luminosity even though it is a very hot star. According to the Stefan–Boltzmann law, luminous stars must be either hot or large. The second case applies to giant stars. Arcturus, for example, is in the phase of hydrogen shell burning and evolves along the red-giant branch. With 25 solar radii, its luminosity is roughly 200 times the solar luminosity. At a distance of only 11.3 pc, it appears as one of the brightest stars on the sky. Betelgeuse is a red supergiant with an enormous diameter of the order of a billion km, but relatively low effective temperature (thus the red color; see the parameters defined in our `stars` dictionary). In the solar system, Betelgeuse would extend far beyond the orbit of Mars, almost to Jupiter. It reaches the luminosity of a hundred thousand Sun-like stars.

3.1.2   Planck Spectrum

The energy spectrum of a black body of temperature T is given by the Planck function

$$B_\lambda(T) = \frac{2hc^2}{\lambda^5} \frac{1}{\exp(hc/\lambda kT) - 1}, \tag{3.3}$$

where λ is the wavelength, h is Planck's constant, k the Boltzmann constant and c the speed of light. The differential $B_\lambda(T) \cos\theta \, d\Omega \, d\lambda$ is the energy emitted per unit surface area and unit time under an angle θ to the surface normal into the solid angle $d\Omega = \sin\theta \, d\theta \, d\phi$ (in spherical coordinates) with wavelengths ranging from λ

---

[5]The radius of Sirius B is 5800 km; see [6].

to $\lambda + d\lambda$. Integration over all spatial directions and wavelengths yields the Stefan–Boltzmann law for the energy flux[6]:

$$F \equiv \pi \int_0^\infty B_\lambda(T)d\lambda = \sigma T^4 . \tag{3.4}$$

We continue our discussion with producing plots of the Planck spectrum for the stars discussed in Sect. 3.1.1 and, additionally, the Sun. The first step is, of course, the definition of a Python function `planck_spectrum()` to compute $B_\lambda(T)$. To avoid numbering of code lines over too many pages, we reset the line number to 1 here. However, be aware that the code listed below nevertheless depends on definitions from above, for example, the `stars` dictionary.

```
1  import numpy as np
2  from scipy.constants import h,c,k
3
4  def planck_spectrum(wavelength, T):
5      """
6      function computes Planck spectrum of a black body
7
8      args: numpy arrays
9            wavelength - wavelength in m
10           T - temperature in K
11
12     returns: intensity in W/m^2/m/sr
13     """
14     return 2*h*c**2 / (wavelength**5 *
15                       (np.exp(h*c/(wavelength*k*T)) - 1))
```

The expression in the function body corresponds to formula (3.3). We use the exponential function from the NumPy library, which allows us to call `planck_spectrum()` with arrays as actual arguments. This is helpful for producing a plot of the Planck spectrum from an array of wavelengths. Moreover, physical constants are imported from `scipy.constants`. Consequently, we do not make use of dimensional quantities here.

The next step is to generate an array of temperatures for the different stars and and a wavelength grid to plot the corresponding spectra. To that end, we collect the values associated with the key `'Teff'` in our dictionary. A wavelength grid with a given number of points between minimum and maximum values is readily generated with `np.linspace()` introduced in Sect. 2.3.

---

[6]Planck's formula thus resolved a puzzle of classical physics which is known as ultraviolet catastrophe (the integrated intensity of black body radiation in a cavity would diverge according to the laws of classical physics). It is not only of fundamental importance in astrophysics but also paved the way for the development of quantum mechanics when Planck realized in 1900 that black body radiation of frequency $\nu = c/\lambda$ can be emitted only in quanta of energy $h\nu$. Owing to the exponential factor in Eq. (3.3), emission of radiation with frequency higher than kT/h is suppressed. This resolves the ultraviolet catastrophe.

```
16  # initialize array for temperatures
17  T_sample = np.zeros(len(stars) + 1)
18
19  # iterate over stellar temperatures in dictionary
20  for i,key in enumerate(stars):
21      T_sample[i] = stars[key]['Teff'].value
22  # add effective temperature of Sun as last element
23  T_sample[-1] = 5778
24
25  # sort temperatures
26  T_sample = np.sort(T_sample)
27
28  # uniformly spaced grid of wavenumbers
29  n = 1000
30  lambda_max = 2e-6
31  wavelength = np.linspace(lambda_max/n, lambda_max, n)
```

Remember that the loop variable of a `for` loop through a dictionary runs through the keys of the dictionary (see Sect. 2.1.2). Here, `key` runs through the names of the stars. By using `enumerate()`, we also have an index i for the corresponding elements of the array `T_sample`, which is initialized as an array of zeros in line 17. The array length is given by the length of the dictionary, i.e. the number of stars, plus one element for the Sun. After the loop, the NumPy function `sort()` sorts the temperatures in ascending order. The minimum wavelength is set to `lambda_max/n` corresponding to the subdivision of the interval $[0, \lambda_{max}]$ with $\lambda_{max} = 2 \times 10^{-6}$ m into 1000 equidistant steps. Although the Planck function is mathematically well defined even for $\lambda = 0$, the numerical evaluation poses a problem because Python computes the different factors and exponents occurring in our definition of the function `planck_spectrum()` separately (you can check this by calling `planck_spectrum()` for wavelength zero). For this reason, zero is excluded.

The following code plots the Planck spectrum for the different temperatures using a color scheme that mimics the appearance of stars (or any black body of given temperature) to the human eye. To make use of this scheme, you need to import the function `convert_K_to_RGB()` from a little module that is not shipped with common packages such as NumPy, but is shared as GitHub gist on the web.[7] The colors are based on the widely used RGB color model to represent colors in computer graphics. A particular RGB color is defined by three (integer) values ranging from 0 to 255 to specify the relative contributions of red, green, and blue.[8] White corresponds to (255, 255, 255), black to (0, 0, 0), pure red to (255, 0, 0), etc. This model allows for the composition of any color shade.

---

[7]Download the module from gist.github.com/petrklus/b1f427accdf7438606a6 and place the file in the directory with your source code or notebook.

[8]Integers in this range correspond to 8 bit values.

```
32  import matplotlib.pyplot as plt
33  from rgb_to_kelvin import convert_K_to_RGB
34  %matplotlib inline
35
36  plt.figure(figsize=(6,4), dpi=100)
37
38  for T in T_sample:
39      # get RGB color corresponding to temperature
40      color = tuple([val/255 for val in convert_K_to_RGB(T)])
41
42      # plot Planck spectrum (wavelength in nm,
43      # intensity in kW/m^2/nm/sr)
44      plt.semilogy(1e9*wavelength, \
45                   1e-12*planck_spectrum(wavelength, T), \
46                   color=color, label="{:.0f} K".format(T))
47
48  plt.xlabel("$\lambda$ [nm]")
49  plt.xlim(0,1e9*lambda_max)
50  plt.ylabel("$B_\lambda(T) $" + \
51             "[$\mathrm{kW\,m^{-2}\,nm^{-1}\, sr^{-1}}$]")
52  plt.ylim(0.1,5e4)
53  plt.legend(loc="upper right", fontsize=8)
54  plt.savefig("planck_spectrum.pdf"
```

We use `semilogy()` from `pyplot` for semi-logarithmic scaling because the Planck functions for the different stars vary over several orders of magnitude. While `convert_K_to_RGB()` returns three integers, a tuple of three floats in the range from 0 to 1 is expected as `color` argument in line 46. As an example of somewhat more fancy Python programming, the conversion is made by an inline loop through the return values (see line 40), and the resulting RGB values are then converted into a tuple via the built-in function `tuple()`. To show formatted mathematical symbols in axes labels, it is possible to render text with LaTeX. We do not cover LaTeX here,[9] but examples can be seen in lines 48 and 50–51. For instance, the greek letter λ in the x-axis label is rendered with the LaTeX code `$\lambda$`, and $B_\lambda(T)$ for the y axis is coded as `$B_\lambda(T)$`.

The resulting spectra are shown in Fig. 3.1. The temperatures are indicated in the legend. The color changes from orange at temperatures around 3000 K to bluish above 10000 K. Our Sun with an effective temperature of 5778 K appears nearly white with a slightly reddish tinge, in agreement with our perception.[10] You can also see that the curves for different temperatures do not intersect, i.e. $B_\lambda(T_2) > B_\lambda(T_1)$ for $T_2 > T_1$. Equation (3.4) implies that the radiative flux F (the total area under the Planck spectrum) is larger for higher temperatures. In other words, a hotter surface emits more energy. Moreover, Fig. 3.1 shows that the peak of the spectrum shifts with increasing temperature to shorter wavelengths. This is, of course, related to the

---

[9]There are plenty of tutorials on the web, for example,
www.overleaf.com/learn/latex/Learn_LaTeX_in_30_minutes.

[10]Of course, our definition of "white" light originates from the adaption of human eyes to daylight.

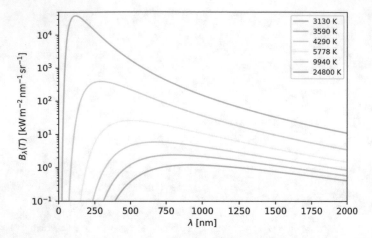

Fig. 3.1 Planck spectra for effective temperatures of different stars

overall color (a larger fraction of short wavelengths corresponds to bluer color). The hottest star in our sample, which is the white dwarf Sirius B, even emits most of its radiation in the UV (below 400 nm).

The position of the maximum of $B_\lambda(T)$ is described by Wien's displacement law:

$$\lambda_{\max} = \frac{b}{T},$$
(3.5)

where $b \approx hc/(4.965114\,k) = 0.002897772$ m K. To determine the maximum of $B_\lambda(T)$ for a given temperature, we need to find the roots of the first derivative with respect to λ:

$$\frac{\partial B_\lambda(T)}{\partial \lambda} = 0\,,$$

which implies

$$\frac{hc}{\lambda kT} \frac{\exp(hc/\lambda kT)}{\exp(hc/\lambda kT) - 1} - 5 = 0\,.$$

This is a transcendent equation that can be solved only numerically. The equation can be simplified with the substitution $x = hc/\lambda kT$. After rearranging the terms, the following equation in terms of x is obtained:

$$f(x) := (x - 5)e^x + 5 = 0\,.$$
(3.6)

To match Wien's displacement law (3.5), the solution should be $x \approx 4.965114$.

We apply an algorithm which is known as bisection method to find the roots of the function $f(x)$ defined by the left-hand side of Eq. (3.6). The idea is really simple. It is known from calculus that a continuous real function $f(x)$[11] has at least one root $x \in [a, b]$ if $f(a)f(b) < 0$, i.e. the function has opposite sings at the endpoints of the interval $[a, b]$ and, consequently, crosses zero somewhere within the interval. This property can be used to find approximate solutions of $f(x) = 0$ by splitting the interval $[a, b]$ at its midpoint $x = (a + b)/2$ into two subintervals $[a, x]$ and $[x, b]$ (thus the name bisection) and checking which subinterval in turn contains a root. This procedure can be repeated iteratively. Since the interval length $|a - b|$ decreases by a factor after each iteration, the iteration ends once $|a - b|$ becomes smaller than a given tolerance ϵ, corresponding to the desired accuracy of the numerical approximation. This means that the final approximation x does not deviate by more than ϵ from the exact root.

The bisection method is implemented in the following Python function. As you know already from many examples, a Python function is generally not equivalent to a function in the mathematical sense, such as `planck_spectrum()`. It is similar to what is called a subprogram or subroutine in other programming languages, providing a well defined interface to an algorithm with various elements of input (the arguments of the function) and output (data returned by the function):

```
 1  def root_bisection(f, a, b, eps=1e-3, verbose=False):
 2      """
 3      bisection algorithm for finding the root of a function f(x)
 4
 5      args: f - function f(x)
 6            a - left endpoint of start interval
 7            b - right endpoint of start interval
 8            eps - tolerance
 9            verbose - print additional information if true
10
11      returns: estimate of x for which f(x) = 0
12      """
13      i = 0 # counter of number of iterations
14
15      # iterate while separation of endpoints
16      # is greater than tolerance
17      while abs(b-a) > eps:
18
19          if verbose:
20              print(f"{a:6.3f} {f(a):10.3e}",
21                    f"{b:6.3f} {f(b):10.3e}")
22
23          # new midpoint
24          x = 0.5*(a+b)
25
```

---

[11]Continuous means that the function has no jumps or poles.

```
26          # check if function crosses zero in left subinterval
27          # and reset endpoint
28          if f(a)*f(x) < 0:
29              b = x
30          else:
31              a = x
32
33          # increment counter
34          i += 1
35
36      print("tolerance reached after {:d} iterations".format(i))
37      print("deviation: f(x) = {:.3e}".format(f(x)))
38
39      return x
```

Let us first look at the different arguments. The first argument is expected to be a function (more precisely, the name of a function) that can be called inside root_bisection(). As we shall see shortly, it is an extremely useful feature of Python that functions can be passed as arguments to other functions.[12] The following two arguments, a and b, are simple variables for the two endpoints of the initial interval. The remaining arguments are optional, which is indicated by assigning default values in the definition of the function. This allows the user to omit actual arguments in place of eps and verbose when calling root_bisection(). Unless an optional argument is explicitly specified, its default value will be assumed, for example, 1e-3 for the tolerance eps and False for verbose.

The bisection algorithm is implemented in the **while** loop starting at line 17. The loop continues as long as **abs**(b-a), where **abs**() returns the absolute value, is larger than the tolerance eps. In lines 28–29, the subinterval $[a, x]$ is selected if the function has opposite signs at the endpoint a and the midpoint $x = (a + b)/2$. In this case, the value of x defined in line 24 is assigned to the variable b for the next iteration (recall the difference between an assignment and equality in the mathematical sense). Otherwise variable a is set equal to x, corresponding to the subinterval $[x, b]$ (lines 30–31). Once the tolerance is reached, the loop terminates and the latest midpoint is returned as final approximation.

Before applying root_bisection() to the Planck spectrum, it is a good idea to test it for a simple function with known roots. Let us try the quadratic polynomial

$$f(x) = x^2 - x - 2$$

---

[12]Like variables, a function is just an object in Python. From this point of view, there is actually nothing unusual about passing functions as arguments to other functions.

with roots $x_1 = -1$ and $x_2 = 2$. In Python, we can define this function as

```
40   def quadratic(x):
41       return x**2 - x - 2
```

Using interactive Python, the call

```
42   root_bisection(quadratic, 0, 5, verbose=True)
```

produces the following output (the return value is printed automatically in an output cell):

```
0.000 -2.000e+00   5.000   1.800e+01
0.000 -2.000e+00   2.500   1.750e+00
1.250 -1.688e+00   2.500   1.750e+00
1.875 -3.594e-01   2.500   1.750e+00
1.875 -3.594e-01   2.188   5.977e-01
1.875 -3.594e-01   2.031   9.473e-02
1.953 -1.384e-01   2.031   9.473e-02
1.992 -2.338e-02   2.031   9.473e-02
1.992 -2.338e-02   2.012   3.529e-02
1.992 -2.338e-02   2.002   5.863e-03
1.997 -8.780e-03   2.002   5.863e-03
2.000 -1.465e-03   2.002   5.863e-03
2.000 -1.465e-03   2.001   2.198e-03
tolerance reached after 13 iterations
deviation: f(x) = 3.662e-04


2.0001220703125
```

By calling root_bisection() with the identifier quadratic as first argument, the bisection method is applied to our test function quadratic() defined in lines 40–41 in place of the generic function f(). In the same way, any other function with an arbitrary name can be used. This is why specifying the function for which we want to compute a root as an argument of root_bisection() is advantageous. While the optional argument eps is skipped in the example above, setting the verbosity flag verbose=True allows us to see how the endpoints a and b of the interval change with each iteration. The Boolean variable verbose controls additional output in lines 20–21, which comes in useful if something unexpected happens and one needs to understand in detail how the code operates. The argument name verbose is also called a keyword in this context. As expected, the function always has opposite signs at the endpoints and the root $x_2 = 2$ is found within a tolerance of 10^{-3} after 13 bisections of the initial interval [0, 5]. So far, everything works fine.

However, try

```
43 | root_bisection(quadratic, -2, 0, verbose=True)
```

The output is

```
-2.000   4.000e+00   0.000 -2.000e+00
-1.000   0.000e+00   0.000 -2.000e+00
-0.500 -1.250e+00   0.000 -2.000e+00
-0.250 -1.688e+00   0.000 -2.000e+00
-0.125 -1.859e+00   0.000 -2.000e+00
-0.062 -1.934e+00   0.000 -2.000e+00
-0.031 -1.968e+00   0.000 -2.000e+00
-0.016 -1.984e+00   0.000 -2.000e+00
-0.008 -1.992e+00   0.000 -2.000e+00
-0.004 -1.996e+00   0.000 -2.000e+00
-0.002 -1.998e+00   0.000 -2.000e+00
tolerance reached after 11 iterations
deviation: f(x) = -1.999e+00
```

```
-0.0009765625
```

Obviously, the algorithm does not converge to the solution $x_1 = -1$ for the start interval $[-2, 0]$.[13] The problem can be spotted in the second line of output (i.e. the second iteration). At the left endpoint $a = -1$, we have $f(a) = 0$. However, `root_bisection()` tests whether $f(a)f(x) < 0$ in line 28. Since the product of function values is zero in this case, the **else** clause is entered and the left endpoint is set to the midpoint $x = -0.5$, as can be seen in the next line of output. As a result, our current implementation of the algorithm misses the exact solution and ends up in an interval that does not contain any root at all (the signs at both endpoints are negative) and erroneously concludes that the variable x contains an approximate solution after the interval has shrunk enough.

Fortunately, this can be easily fixed by testing whether $f(x)$ happens to be zero for some given x. The improved version of `root_bisection()` is listed in the following (the comment explaining the function is omitted).

```
1 | def root_bisection(f, a, b, eps=1e-3, verbose=False):
2 |     i = 0 # counter of number of iterations
3 |
4 |     # iterate while separation of endpoints
5 |     # is greater than tolerance
6 |     while abs(b-a) > eps:
7 |
8 |         if verbose:
```

---

[13]If we had chosen e.g. $[-5, 0]$, this problem would not occur. But the algorithm must yield a correct answer for every choice that meets the criteria for applying the bisection method.

```
 9              print(f"{a:6.3f} {f(a):10.3e}",
10                    f"{b:6.3f} {f(b):10.3e}")
11
12          # new midpoint
13          x = 0.5*(a+b)
14
15          # check if function crosses zero in left subinterval
16          # and reset endpoint unless x is exact solution
17          if f(x) == 0:
18              print("found exact solution " + \
19                    "after {:d} iteration(s)".format(i+1))
20              return x
21          elif f(a)*f(x) < 0:
22              b = x
23          else:
24              a = x
25
26          # increment counter
27          i += 1
28
29      print("tolerance reached after {:d} iterations".format(i))
30      print("deviation: f(x) = {:.3e}".format(f(x)))
31
32      return x
```

The keyword `elif` introduces a third case between the `if` and `else` clauses. The `if` statement in line 17 checks if the value of x is an exact root. If this is the case, the current value will be returned immediately, accompanied by a message. As you can see, a `return` statement can occur anywhere in a function, not only at the end. Of course, this makes sense only if the statement is conditional, otherwise the remaining part of the function body would never be executed. If the condition `f(x) == 0` evaluates to `False`, the algorithm checks if the function has opposite signs at the endpoints of the subinterval $[a, x]$ (`elif` statement in line 21) and if that is not case either (`else` statement), the right subinterval $[x, b]$ is selected for the next iteration.

To test the modified algorithm, we first convince ourselves that answer for the initial interval $[0, 5]$ is the same as before:

```
33  root_bisection(quadratic, 0, 5)
```

prints (`verbose` set to `False` by default)

```
tolerance reached after 13 iterations
deviation: f(x) = 3.662e-04

2.0001220703125
```

Now let us see if we obtain the root x_1 for the start interval $[-2, 0]$:

```
34  root_bisection(quadratic, -2, 0)
```

Indeed, our modification works:

```
found exact solution after 1 iteration(s)


-1.0
```

To obtain a more accurate approximation for x_2, we can prescribe a lower tolerance:

```
35  root_bisection(quadratic, 0, 5, 1e-6)
```

In this case, it is not necessary to explicitly write eps=1e-6 in the function call because the actual argument 1e-6 at position four in the argument list corresponds to the fourth formal argument eps in the definition of the function. Arguments that are uniquely identified by their position in the argument list are called *positional* arguments. If an argument is referred by its name rather than position (like verbose=True in the examples above), it is called a *keyword* argument (the keyword being the name of the corresponding formal argument). In particular, an optional argument must be identified by its keyword if it occurs at a different position in the function call.[14] The result is

```
tolerance reached after 23 iterations
deviation: f(x) = -3.576e-07


1.9999998807907104
```

So far, the initial interval $[a, b]$ was chosen such that it contains only one root and the solution is unambiguous. However, what happens if root_bisection() is executed for, say, the interval $[-5, 5]$? The bisection method as implemented above returns only one root, although both x_1 and x_2 are in $[-5, 5]$. It turns out that

```
36  root_bisection(quadratic, -5, 5)
```

converges to x_1:

```
tolerance reached after 14 iterations
deviation: f(x) = 1.099e-03


-1.0003662109375
```

We leave it as an exercise to study in detail how this solution comes about by using the verbose option. Basically, the reason for obtaining x_1 rather than x_2 is that we chose to first check whether $f(a)f(x) < 0$ rather than $f(b)f(x) < 0$. In other

---

[14]In principle, you can also pass non-optional arguments as keyword arguments at arbitrary positions, but this is unnecessary and should be avoided.

words, the answer depends on our specific implementation of the bisection method. This is not quite satisfactory.

Is it possible at all to implement a more robust algorithm that does not depend on any particular choices? The trouble with the bisection method is that there is no unique solution for any given interval $[a, b]$ for which $f(a)f(b) < 0$. If the function happens to have multiple roots, the bisection method will converge to any of those roots, but it is unclear to which one. The only way out would be to make sure that all of them are found. This can indeed be accomplished by means of a recursive algorithm. In Python, recursion means that a function repeatedly calls itself. As with loops, there must be a condition for termination. Otherwise an infinite chain of function calls would result. In the case of the bisection method, this condition is $f(a)f(b) \geq 0$ or $|b - a| < \epsilon$ (where "or" is understood as the inclusive "or" from logic), i.e. the function stops to call itself once the endpoint values for any given interval have equal signs or at least one of the two values is zero or the tolerance for the interval length is reached.

The following listing shows how to find multiple roots through recursion.

```
1   def root_bisection(f, a, b, roots, eps=1e-3):
2       """
3       recursive bisection algorithm for finding multiple roots
4       of a function f(x)
5
6       args: f - function f(x)
7             a - left endpoint of start interval
8             b - right endpoint of start interval
9             roots - numpy array of roots
10            eps - tolerance
11
12      returns: estimate of x for which f(x) = 0
13      """
14      # midpoint
15      x = 0.5*(a+b)
16
17      # break recursion if x is an exact solution
18      if f(x) == 0:
19          roots = np.append(roots, x)
20          print("found {:d}. solution (exact)".
21              format(len(roots)))
22      # break recursion if tolerance is reached
23      elif abs(h-a) <- eps:
24          roots = np.append(roots, x)
25          print("found {:d}. solution,".format(len(roots)),
26              "deviation f(x) = {:6e}".format(f(x)))
27      # continue recursion if function crosses zero
28      # in any subinterval
29      else:
30          if f(a)*f(x) <= 0:
31              roots = root_bisection(f, a, x, roots, eps)
32          if f(x)*f(b) <= 0:
```

```
33                    roots = root_bisection(f, x, b, roots, eps)
34
35      return roots
```

To understand the recursive algorithm, consider an arbitrary interval $[a, b]$ with $f(a)f(b) < 0$ and midpoint $x = (a + b)/2$. We can distinguish the following four cases:

1. If $f(x) = 0$, then x is an exact solution.
2. If $|a - b| \leq \epsilon$, then x is an approximate solution within the prescribed tolerance.
3. If $|a - b| > \epsilon$ and $f(a)f(x) < 0$, then there is at least one root in the open interval $x \in]a, x[$.
4. If $|a - b| > \epsilon$ and $f(x)f(b) < 0$, then there is at least one root in the open interval $x \in]x, b[$.[15]

The nested control structure beginning in line 18 implements the corresponding courses of action. First the code checks if x is an exact solution or if the tolerance is already reached (cases 1. and 2.). In both cases, the value of the local variable x (i.e. the exact or approximate root) is added to the array roots before the algorithm terminates. We utilize the NumPy function append() to add a new element to an existing array. The resulting array is then re-assigned to roots. Only if x is neither an exact nor an approximate root of sufficient accuracy, root_bisection() will be called recursively for one or both subintervals (lines 30–33). This is different from the iterative algorithms discussed at the beginning of this section, where only a single subinterval is selected in each iteration step. Depending on whether only case 3. or 4. or both cases apply, the recursive algorithm progresses like the iterative algorithm toward a single root or branches to pursue different roots. The algorithm can branch an arbitrary number of times. Each call of root_bisection() returns an array (see line 35) containing roots appended at deeper recursion levels (the recursion depth increases with the number of recursive functions calls). Since np.append() generates a new object rather than modifying an existing object, the arrays returned by the calls in lines 31 and 33 have to be re-assigned to roots in the namespace of the calling instance. It does not matter that the name roots is used both for the array received as argument and for the array returned by the function. We use the same name only to highlight the recursive nature of the algorithm.

As a first test,

```
36  x0 = root_bisection(quadratic, -2, 0, [])
37  print(x0)
```

should find the solution $x_1 = -1$ as in the example above. This is indeed the case:

```
found 1. solution (exact)
[-1.]
```

---

[15] An open interval $]a, b[$ contains all values $a < x < b$, i.e. the endpoints are excluded.

The value returned by `root_bisection()` is now printed in brackets, which indicates an array. We started with an empty array `[]` (technically speaking, an empty list that is automatically converted into an array) in the initial function call and obtained an array `x0` with a single element, namely -1. This array can in turn be passed as actual argument to another call of `root_bisection()` for a different start interval:

```
38  x0 = root_bisection(quadratic, 0, 5, x0)
39  print(x0)
```

The array returned by the second call of `root_bisection()` is re-assigned to `x0` and now contains both roots:

```
found 2. solution, deviation f(x) = -5.492829e-04
[-1.          1.99981689]
```

Try to figure out how the code determines that this is the second solution (as indicated by "`found 2. solution`" in the output).

Of course, we would not gain anything compared to the iterative implementation if it were not for the possibility of computing all roots at once. For example, with

```
40  x0 = root_bisection(quadratic, -5, 5, [])
41  print(x0)
```

we receive the output

```
found 1. solution, deviation f(x) = 1.831092e-04
found 2. solution, deviation f(x) = -5.492829e-04
[-1.00006104  1.99981689]
```

In this case, the first root is also approximate because the bisection starts with the left endpoint $a = -5$ instead of -2 and the resulting midpoints do not hit the exact solution $x_1 = -1$. As a small exercise, you may modify the recursive function to print detailed information to see how the intervals change with increasing recursion depth.

After having carried out various tests, we are now prepared to numerically solve Eq. (3.6). First we need to define the function $f(x)$ corresponding to the left-hand side of the equation:

```
42  dof f(x).
43      return (x-5)*np.exp(x) + 5
```

Since the solution of $f(x) = 0$ is some number $x \sim 1$ (for large x, exponential growth would rapidly dominate), let us try the start interval $[0, 10]$:

```
44  x0 = root_bisection(f, 0, 10, [])
45  print(x0)
```

The recursive bisection method finds two solutions

```
found 1. solution, deviation f(x) = -1.220843e-03
found 2. solution, deviation f(x) = -2.896551e-02
[3.05175781e-04 4.96490479e+00]
```

The first solution is actually the endpoint $a = 0$ for which $f(a) = 0$. The second solution ≈ 4.965 is approximately the solution we know from Wien's law (you might want to plot $f(x)$ using `pyplot` to get an impression of the shape of the function). You can improve the accuracy by specifying a smaller tolerance `eps` and check the agreement with the value cited above. If you compare the conditions for zero-crossings in the iterative and recursive implementations, you might wonder why the operator < is used in the former and <= (smaller or equal to) in the latter. Replace <= by < in the checks of both subintervals in the recursive function and see what happens if you repeat the execution for the start interval [0, 10]. The output will change drastically. Try to figure out why.[16]

The workflow with several cycles of programming and testing is typical for the implementation of a numerical method. The first implementation never works perfectly. This is why simple test cases for which you know the answer beforehand are so important to validate your code. Frequently, you will notice problems you have not thought of in the beginning and sometimes you come to realize there is an altogether better solution.

Exercises

3.1 Extend the function `stellar_parameters()` to add stellar masses to the dictionary `stars`. For the stars discussed in Sect. 3.1.1, the masses in units of the solar mass are 0.144, 2.06, 1.02, 1.1, and 12 (from Bernard's Star to Betelgeuse). Moreover, write a Python function that calculates the radius for given luminosity and effective temperature. Use this function to add Aldebaran ($L = 4.4 \times 10^2 L_\odot$, $T_{\text{eff}} = 3.9 \times 10^3$ K, $M = 1.2 M_\odot$) and Bellatrix ($L = 9.21 \times 10^2 L_\odot$, $T_{\text{eff}} = 2.2 \times 10^4$ K, $M = 8.6 M_\odot$) to the dictionary and carry out the following tasks.

(a) Print a table of the stellar parameters M, R, T_{eff}, and L aligned in columns.
(b) Plot the luminosities versus the effective temperatures in a double-logarithmic diagram using different markers for main-sequence stars, white dwarfs, and red giants. In other words, produce a Herztsprung–Russell diagram for the stars in the dictionary.
(c) Produce a plot of luminosity vs mass in double-logarithmic scaling. Which type of relation is suggested by the data points?

---

[16]The recursive algorithm is still not foolproof though. For example, if a start interval is chosen such that the initial midpoint happens to be a root, recursion will stop right at the beginning and other zeros inside the start interval will be missed. Fixing such a special case without entailing other problems is somewhat tricky.

3.2 In astronomy, the observed brightness of an object on the sky is specified on a logarithmic scale that is based on flux ratios [3, Chap. 4][17]:

$$m = M - 2.5 \log_{10} \left(\frac{F}{F_0} \right) .$$

The convention is to define $F_0 = L/4\pi r_0^2$ as the radiative flux at distance $r_0 = 10$ pc. While the absolute magnitude M is a constant parameter for a star of given luminosity L, the star's apparent magnitude m depends on its distance r The relation $F \propto 1/r^2$ for the radiative flux implies

$$m - M = 5 \log_{10} \left(\frac{r}{10 \text{ pc}} \right) .$$

Hence, the distance of the star can be determined if both the apparent and absolute magnitude are known. However, extinction due to interstellar dust modifies this relation:

$$m - M = 5 \log_{10} \left(\frac{r}{10 \text{ pc}} \right) + kr , \tag{3.7}$$

Although the extinction varies significantly within the Galaxy, the mean value $k = 2 \times 10^{-3}$ pc^{-1} can be assumed for the extinction per unit distance [3, Sect. 16.1].

Compute and plot r in units of pc for B0 main sequence stars with and absolute magnitude $M = -4.0$ and apparent magnitudes m in the range from -4.0 to 6.0 in the visual band.[18] How are the distances affected by extinction? To answer this question, you will need to solve Eq. (3.7) numerically for each data point. To be able to plot a graph, it is advisable to create an array of closely spaced magnitudes.

3.2 Physics of Stellar Atmospheres

To a first approximation, the radiation of stars is black body radiation. However, observed stellar spectra deviate from the Planck function. The most important effect is the absorption of radiation from deeper and hotter layers of a star by cooler gas in layers further outside. The outermost layers which shape the spectrum of the visible radiation of a star are called stellar atmosphere.

Although hydrogen is the most abundant constituent, other atoms and ions play an important role too, especially in stars of low effective temperature. For a more detailed discussion of the physics, see Chaps. 8 and 9 in [4]. As a consequence, modeling the transfer of radiation in stellar atmospheres is extremely complex and requires

---

[17]The factor -2.5 is of ancient origin and dates back to the scale introduced by the Greek astronomer Hipparchus to classify stars that are visible by eye.

[18]Magnitudes are usually measured for a particular wavelength band, see Sect. 4.3 in [3].

numerical computations using databases for a huge number of line transitions for a variety of atoms, ions, and molecules. You will see an application of such model atmospheres in Sect. 5.5.2.

In the following, we will consider some basic aspects of the physical processes in stellar atmospheres. An important source of absorption in stellar atmospheres are transitions from lower to higher energy levels (also known as bound-bound transitions). For hydrogen, the energy difference between levels defined by the principal quantum numbers n_1 and $n_2 > n_1$ is given by

$$\Delta E = -13.6 \text{ eV} \left(\frac{1}{n_2} - \frac{1}{n_1} \right) . \tag{3.8}$$

A photon can be absorbed if its energy matches the energy difference between the two levels:

$$\Delta E = \frac{hc}{\lambda} \tag{3.9}$$

where h is Planck's constant, c is the speed of light, and λ is the wavelength of the photon. For a hydrogen atom in the ground state, $n_1 = 1$, the wavelength associated with the transition to the next higher level, $n_2 = 2$, is $\lambda = 121.6$ nm and even shorter for higher levels. Emission or absorption lines at these wavelengths, which are all ultraviolet, are called the Lyman series. One might expect that the Lyman series can be seen in the light emitted by hot stars. However, the fraction of hydrogen atoms in the ground state rapidly decreases with increasing temperature. It turns out that transitions from the first excited state, $n_1 = 2$, to levels $n_2 > 2$ give rise to prominent absorption lines in stellar spectra, which are known as Balmer lines. To understand how this comes about, we need to apply two fundamental equations from statistical physics, namely the Boltzmann and Saha equations, to compute the fraction of atoms in excited states and the fraction that is ionized at a given temperature.

3.2.1 Thermal Excitation and Ionization

Collisions between atoms excite some of them into a higher energy state, while others lose energy. The balance between these processes is described by the Boltzmann distribution. If the gas is in thermal equilibrium, the ratio of the occupation numbers N_2 and N_1 of levels n_2 and n_1, respectively, is given by

$$\frac{N_2}{N_1} = \frac{g_2}{g_1} e^{-(E_2 - E_1)/kT} \tag{3.10}$$

where g_1 and g_2 are the statistical weights of the energy levels (i.e. the number of possible quantum states with the same energy) and T is the temperature of the gas. For the hydrogen atom, the nth energy level is degenerate with weight $g_n = 2n^2$ (the energy of a state is independent of the spin and the angular momentum quantum numbers).

As an example, let us compute N_2/N_1 for the first excited state of hydrogen for the stars defined in the dictionary `stars` in Sect. 3.1.1:

```
1   import numpy as np
2   from scipy.constants import k,physical_constants
3
4   # ionization energy of hydrogen
5   chi = physical_constants['Rydberg constant times hc in J'][0]
6
7   # energy levels
8   n1, n2 = 1, 2
9
10  print("T [K]   N2/N1")
11  for T in T_sample:
12      print("{:5.0f}   {:.3e}".format(T,
13          (n2/n1)**2 * np.exp(chi*(1/n2**2 - 1/n1**2)/(k*T))))
```

The occupation numbers of the first excited state ($n_2 = 2$) relative to the ground state ($n_1 = 1$) are printed for the effective temperatures in the array `T_sample` (see Sect. 3.1.2). The Boltzmann equation (3.10) is implemented inline as an expression in terms of the variables `n1`, `n2`, and `Teff`. Line 5 defines the ionization energy $\chi \equiv E_\infty - E_1$ in terms of the Rydberg constant R:

$$\chi = hcR = 13.6 \text{ eV}. \tag{3.11}$$

This allows us to express the energy difference ΔE between the two states in Boltzmann's equation as

$$\Delta E = -\chi \left(\frac{1}{n_2} - \frac{1}{n_1} \right). \tag{3.12}$$

The value of χ in SI units (J) is available in SciPy's `physical_constants` dictionary, which is imported in line 2. This dictionary allows us to conveniently reference physical constants via keywords. In the case of χ, the key expresses formula (3.11) in words. Since each item in `physical_constants` is a tuple containing the numerical value, unit, and precision of a constant, we need to assign the first element of the tuple to the variable `chi`.

The code listed produces the output

```
T [K]   N2/N1
 3130   1.485e-16
 3590   1.892e-14
 4290   4.115e-12
 5778   5.030e-09
 9940   2.681e-05
24800   3.376e-02
```

While the fraction of hydrogen in the first excited state is very low for cooler stars and the Sun, it increases rapidly toward the hot end. An easy calculation shows that $N_2 = N_1$ is reached at a temperature of 8.54×10^4 K. This is higher than the effective temperature of even the hottest stars of class O. Transitions from the first excited state thus should become ever more important as the effective temperature increases. However, this is not what is observed: the strongest Balmer absorption lines are seen in the spectra of stars of spectral class A, with effective temperatures below 10000 K.

The reason for this is the ionization of hydrogen. Once a hydrogen atom is stripped of its electron, there are no transitions between energy levels. The temperature-dependent fraction of ionized hydrogen (HII) can be computed using the Saha equation:

$$\frac{N_{\mathrm{II}}}{N_{\mathrm{I}}} = \frac{2kT Z_{\mathrm{II}}}{P_{\mathrm{e}} Z_{\mathrm{I}}} \left(\frac{2\pi m_{\mathrm{e}} kT}{h^2}\right)^{3/2} \mathrm{e}^{-\chi/kT} . \tag{3.13}$$

Similar to the Boltzmann equation (3.10), the ratio $N_{\mathrm{II}}/N_{\mathrm{I}}$ is dominated by the exponential factor $\mathrm{e}^{-\chi/kT}$ for low temperature. Here, χ is the ionization energy defined by Eq. (3.11). An additional parameter is the pressure of free electrons, P_{e} (i.e. electrons that are not bound to atoms). The factor kT/P_{e} equals the inverse number density of free electrons. If there are more free electrons per unit volume, recombinations will become more frequent and the number of ions decreases. Apart from the Boltzmann and Planck constants, the electron mass m_{e} enters the equation. The partition function Z is the effective statistical weight of an atom or ion. It is obtained by summing over all possible states with relative weights given by the Boltzmann distribution (3.10):

$$Z = g_1 \left(1 + \sum_{n=2}^{\infty} \frac{g_n}{g_1} \mathrm{e}^{-(E_n - E_1)/kT}\right) . \tag{3.14}$$

Since we consider a regime of temperatures for which the thermal energy kT is small compared to the energy difference $E_n - E_1$ between the ground state and higher energy levels (in other words, most of the hydrogen is in its ground state), the approximation $Z_{\mathrm{I}} \simeq g_1 = 2$ can be used. This is consistent with the values of N_2/N_1 computed above. The partition function $Z_{\mathrm{II}} = 1$ because a hydrogen ion has no electron left that could occupy different energy levels.

The Saha equation is put into a Python function to evaluate $N_{\mathrm{II}}/N_{\mathrm{I}}$ for given temperature and electron pressure:

```
14  def HII_frac(T, P_e):
15      """
16      computes fraction of ionized hydrogen
17      using the Saha equation
18
19      args: T - temperature in K
20            P_e - electron pressure in Pa
21
22      returns: HII fraction
```

```
23        """
24        E_therm = k*T
25
26        return (E_therm/P_e) * \
27              (2*np.pi*m_e*E_therm/h**2)**(3/2) * \
28              np.exp(-chi/E_therm)
```

While the local variable `E_therm` is used for the thermal energy kT of the gas, the constant ionization energy is defined by the variable `chi` in the global namespace (see line 5 above).

To estimate the strength of Balmer lines, we need to compute the number of neutral hydrogen atoms (HI) in the first excited state relative to all hydrogen atoms and ions:

$$\frac{N_2}{N_\mathrm{I} + N_\mathrm{II}} \simeq \frac{N_2}{N_1 + N_2} \frac{N_\mathrm{I}}{N_\mathrm{I} + N_\mathrm{II}}$$
$$= \frac{N_2/N_1}{1 + N_2/N_1} \frac{1}{1 + N_\mathrm{II}/N_\mathrm{I}} ,$$

(3.15)

where we used the approximation $N_\mathrm{I} \simeq N_1 + N_2$ (fraction of hydrogen in higher excited states is negligible). The fractions N_2/N_1 and $N_\mathrm{II}/N_\mathrm{I}$ can be computed using Eqs. (3.10) and (3.13), respectively.

The electron pressure in stellar atmosphere P_e ranges from about 0.1 to 100 Pa, where the lower bound applies to cool stars. The following Python code computes and plots $N_2/(N_\mathrm{I} + N_\mathrm{II})$ as function of temperature, assuming $P_\mathrm{e} \approx 20$ Pa (200 dyne cm^{-2} in the cgs system) as representative value for the electron pressure.

```
29  import matplotlib.pyplot as plt
30
31  P_e = 20 # electron pressure in Pa
32
33  # temperature in K
34  T_min, T_max = 5000, 25000
35  T = np.arange(T_min, T_max, 100.0)
36
37  # fraction of HI in first excited state
38  HI2_frac = 4*np.exp(-0.75*chi/(k*T))
39
40  # plot fraction of all hydrogen in first excited state
41  plt.figure(figsize=(6,4), dpi=100)
42  plt.plot(T, 1e5*HI2_frac / &
43              ((1 + HI2_frac)*(1 + HII_frac(T, P_e))))
44  plt.xlim(T_min, T_max)
45  plt.xlabel("$T$ [K]")
46  plt.ylim(0, 0.9)
47  plt.ylabel("$10^5\,N_2/N_{\mathrm{I+II}}$")
48  plt.savefig("hydrogen_frac.pdf")
```

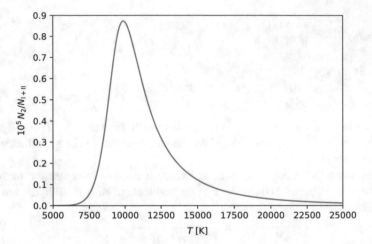

Fig. 3.2 Fraction $N_2/(N_I + N_{II})$ of hydrogen in the first excited state for constant electron pressure $P_e \approx 20$ Pa as function of temperature

In line 38 the Boltzmann equation is applied to compute the fraction N_2/N_1 for an array of temperatures ranging from 5000 to 2.5×10^4 K (variables `T_min` and `T_max` defined in line 34). Expression (3.15) is then used in the call of the plot function in lines 42–43. The resulting fraction is scaled by a factor of 10^5 to obtain numbers of order unity.

The graph in Fig. 3.2 shows that $N_2/(N_I + N_{II})$ peaks at a temperature of about 10000 K in good agreement of the observed strength of Balmer lines in the spectra of stars. This temperature is significantly lower than the temperature $\sim 10^5$ K for which most of the hydrogen atoms would be in an excited state. However at such high temperatures, almost all hydrogen is ionized. As a result, the fraction $N_2/(N_I + N_{II})$ does not exceed $\sim 10^{-5}$ even at the peak. Nevertheless, such a small fraction is sufficient to produce strong Balmer absorption lines in the atmospheres of A stars. The first line in the Balmer series is called Hα line and has a wavelength $\lambda = 656.45$ nm, which is in the red part of the spectrum. Transitions to higher levels ($n_2 > 3$) are seen as absorption lines Hβ, Hγ, etc. at wavelengths ranging from blue to ultraviolet. Observed spectra of stars of different spectral classes can be found in [3, Sect. 9.2] and [4, Sect. 8.1].

So far, we have ignored the dependence on electron pressure. Since high electron pressure suppresses ionization, it appears possible that we overestimated the decline of $N_2/(N_I + N_{II})$ toward high temperature. To investigate the dependence on both parameters (temperature and electron pressure), it is helpful to produce a three-dimensional surface plot:

```
49  from mpl_toolkits.mplot3d import Axes3D
50  from matplotlib.ticker import LinearLocator
51
52  fig = plt.figure(figsize=(6,4), dpi=100)
```

```
53  ax = plt.axes(projection='3d')
54
55  P_min, P_max  = 10, 100
56
57  # create meshgrid
58  # (x-axis: temperature, y-axis: electron pressure)
59  T, P_e = np.meshgrid(np.arange(T_min, T_max, 200.0),
60                        np.arange(P_min, P_max, 1.0))
61
62  # fraction of HI in first excited state
63  HI2_frac = 4*np.exp(-0.75*chi/(k*T))
64
65  # create surface plot
66  surf = ax.plot_surface(T, P_e,
67      1e5*HI2_frac/((1 + HI2_frac)*(1 + HII_frac(T, P_e))),
68      rcount=100, ccount=100,
69      cmap='BuPu', antialiased=False)
70
71  # customize axes
72  ax.set_xlim(T_min, T_max)
73  ax.set_xlabel("$T$ [K]")
74  ax.xaxis.set_major_locator(LinearLocator(5))
75  ax.set_ylim(P_min, P_max)
76  ax.set_ylabel("$P_{\mathrm{e}}$ [Pa]")
77
78  # add color bar for z-axis
79  cbar = fig.colorbar(surf, shrink=0.5, aspect=5)
80  cbar.ax.set_ylabel("$10^5 N_2/N_{\mathrm{I+II}}$")
81
82  plt.savefig("hydrogen_frac_3d.png")
```

Surface plots can be produced with pyplot using the `mplot3d` toolkit. In line 53, a three-dimensional axes object is created (this is based on `Axes3D` imported in line 49). To define the data points from which the surface is rendered, we need a two-dimensional array of points in the xy-plane and the corresponding z-values (in our example, $N_2/(N_I + N_{II})$ as function of T and P_e). The x- and y-coordinates are defined by one-dimensional arrays of temperatures and pressures in steps of 200 K and 1 Pa, respectively. The function `np.meshgrid()`, which you know from Sect. 2.3, generates a meshgrid of coordinate pairs in the xy-plane. The resulting two-dimensional arrays followed by the corresponding values of $N_2/(N_I + N_{II})$ defined by Eq. (3.15) are passed as arguments in the call of `ax.plot_surface()` in lines 66–67. As in the two-dimensional plot above, we scale the z-axis by a factor of 10^5. The optional argument `cmap='BuPu'` in line 69 specifies a colormap,[19] and further arguments control how the surface is rendered and displayed. The coloring

---

[19]See matplotlib.org/3.1.1/tutorials/colors/colormaps.html for further details and available colormaps.

Fig. 3.3 The same quantity
as in Fig. 3.2 shown as
function of temperature and
electron pressure

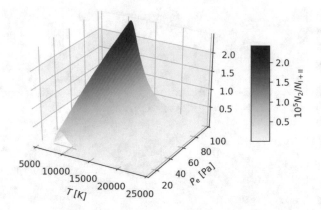

of the surface corresponds to the height in z-direction, i.e. the function value, as indicated by the colorbar created in lines 79–80.

After rendering the surface, further methods are called to customize the axes. For example, `ax.set_xlim()` sets the plot range along the x-axis. The method `xaxis.set_major_locator()` can be used to set the major ticks labeled by numbers along the x-axis. Tick locating and formatting is defined in `matplotlib.ticker` (see line 50). In line 74, the tick locator is informed to use five evenly spaced ticks along the x-axis. This prevents the axis from becoming too crowed by numbers (experiment with the settings to see for yourself how the appearance of the plot is affected). All of this might sound rather complicated, but you will easily get accustomed to making such plots by starting from examples without worrying too much about the gory details.

The graphical output is shown in Fig. 3.3. It turns out that the maximal fraction of excited hydrogen increases with electron pressure (this is expected because ionization is reduced), but the location of the peak is shifted only slightly from lower to higher temperature. Even for the upper bound of P_e in stellar atmospheres (about 100 Pa), the increase is quite moderate (little more than a factor of two compared to Fig. 3.2). Of course, P_e and T are not really independent variables. Conditions in stellar atmospheres will roughly follow a diagonal cut through the surface shown in Fig. 3.3 from low P_e in cool stars toward higher P_e in hot stars. Computing P_e requires a detailed model of the stellar atmosphere. For this purpose, researchers have written elaborate codes.

3.2.2 The Balmer Jump

Balmer lines result from absorbed photons that lift electrons from the first excited state of hydrogen, $n_1 = 2$, to higher energy levels, $n_2 > 2$. If a photon is sufficiently energetic, it can even ionize a hydrogen atom. The condition for ionization from the state $n_1 = 2$ is

$$\frac{hc}{\lambda} \geq \chi_2 = \frac{13.6 \text{ eV}}{2^2} = 3.40 \text{ eV} . \tag{3.16}$$

The corresponding maximal wavelength is 364.7 nm. Like the higher Balmer lines, it is in the ultraviolet part of the spectrum. Since ionizing photons can have any energy above χ_2, ionization will result in a drop in the radiative flux at wavelengths shorter than 364.7 nm rather than a line. This is called the Balmer jump.

To estimate the fraction of photons of sufficient energy to ionize hydrogen for a star of given effective temperature, let us assume that the incoming radiation is black body radiation. From the Planck spectrum (3.3), we can infer the flux below a given wavelength:

$$F_{\lambda \leq \lambda_0} = \pi \int_0^{\lambda_0} \frac{2hc^2}{\lambda^5} \frac{1}{\exp(hc/\lambda kT) - 1} d\lambda . \tag{3.17}$$

Since we know that the total radiative flux integrated over all wavelengths is given by Eq. (3.4), the fraction of photons with wavelength $\lambda \leq \lambda_0$ is given by $F_{\lambda \leq \lambda_0}/F$.

Since the integral in Eq. (3.17) cannot be solved analytically, we apply numerical integration.[20] The definite integral of a function $f(x)$ is the area below the graph of the function for a given interval $x \in [a, b]$. The simplest method of numerical integration is directly based on the notion of the Riemann integral:

$$\int_a^b f(x) dx = \lim_{N \to \infty} \sum_{n=1}^N f(x_{n-1/2}) \Delta x , \tag{3.18}$$

where $\Delta x = (b - a)/N$ is the width of the nth subinterval and $x_{n-1/2} = a + (n - 1/2)\Delta x$ is its midpoint. The sum on the right-hand side means that the area is approximated by N rectangles of height $f(x_n)$ and constant width Δx. If the function meets the basic requirements of Riemann integration (roughly speaking, if it has no poles and does not oscillate within arbitrarily small intervals), the sum converges to the exact solution in the limit $N \to \infty$. In principle, approximations of arbitrarily high precision can be obtained by using a sufficient number N of rectangles. This is called rectangle or midpoint rule.

More efficient methods use shapes that are closer to the segments of the function $f(x)$ in subintervals. For example, the trapezoidal rule is based on a piecewise linear approximation to the function (i.e. linear interpolation between subinterval endpoints). In this case, the integral is approximated by a composite of N trapezoids, as illustrated in Fig. 3.4. The resulting formula for the integral can be readily deduced from the area $(f(x_n - 1) + f(x_n)) \Delta x/2$ of the nth trapezoid:

$$\int_a^b f(x) dx \simeq \left(\frac{1}{2} f(a) + \sum_{n=1}^{N-1} f(x_n) + \frac{1}{2} f(b) \right) \Delta x , \tag{3.19}$$

---

[20]See also [7, Chap. 9] for numerical integration using Python.

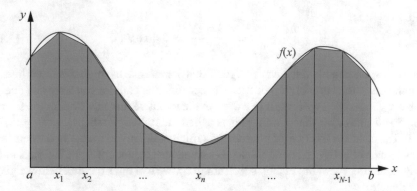

Fig. 3.4 Numerical integration of a function using the trapezoidal rule

where $x_n = a + n\Delta x$ for $0 \leq n \leq N$ are the subinterval endpoints (in particular, $x_0 = a$ and $x_N = b$).

The error of the approximations can be further reduced by quadratic interpolation of function values. This is the underlying idea of Simpson's rule, which can be expressed as

$$\int_a^b f(x)\mathrm{d}x \simeq \left(f(a) + 2 \sum_{k=1}^{N/2-1} f(x_{2k}) + 4 \sum_{k=1}^{N/2} f(x_{2k-1}) + f(b) \right) \frac{\Delta x}{3} . \quad (3.20)$$

Here, the summation index k is mapped to even and odd numbers in the first and second sum, respectively. As a result, we have the terms $f(x_2), f(x_4), \ldots, f(x_{N-2})$ in the first sum and $f(x_1), f(x_3), \ldots, f(x_{N-1})$ in the second sum.

We will now put these methods into practice, starting with the trapezoidal rule. Here is a straightforward implementation of Eq. (3.19) using NumPy arrays:

```
 1  def integr_trapez(f, a, b, n):
 2      """
 3      numerical integration of a function f(x)
 4      using the trapezoidal rule
 5
 6      args: f - function f(x)
 7            a - left endpoint of interval
 8            b - right endpoint of interval
 9            n - number of subintervals
10
11      returns: approximate integral
12      """
13
14      # integration step
15      h = (b - a)/n
```

```
16
17      # endpoints of subintervals between a+h and b-h
18      x = np.linspace(a+h, b-h, n-1)
19
20      return 0.5*h*(f(a) + 2*np.sum(f(x)) + f(b))
```

The subinterval width Δx is commonly denoted by h and called integration step in the numerical context. Subinterval endpoints x_n (excluding x_0 and x_N) are organized in an array that is passed as argument to a generic function f(). We implicitly assume that f() accepts array-like arguments, such as root_bisection() in Sect. 3.1.2. The array of function values returned by f() is in turn passed to the NumPy function sum() to sum up all elements. Thus, the expression np.sum(f(x)) in line 20 corresponds to the sum in Eq. (3.19).

As a simple test case, let us apply the trapezoidal rule to the integral

$$\int_0^{\pi/2} \sin(x)\,\mathrm{d}x = 1$$

We can use the sine function from NumPy:

```
21  print("  n   integr")
22  for n in range(10,60,10):
23      print("{:2d}   {:.6f}".format(n,
24          integr_trapez(np.sin, 0, np.pi/2, n)))
```

The output demonstrates that our implementation works and $N = 20$ subdivisions of the integration interval are sufficient to reduce the numerical error below 10^{-3}:

```
 n   integr
10   0.997943
20   0.999486
30   0.999772
40   0.999871
50   0.999918
```

It is also clear that increasing N further results only in minor improvements of the accuracy of the result.

Before continuing with Simpson's rule, imagine for a moment you had chosen zero as start value of range in line 22. This would have thrown a ZeroDivisionError in the first iteration and the remainder of the loop would not have been executed (try it). The error is caused by the division through n in line 15 in the body of integr_trapez(). A detailed error message will point you to this line and you would probably be able to readily fix the problem. To prevent a program from crashing in the first place, Python offers a mechanism to continue with execution even in the case of an error. This is known as exception handling. In Python, exception handling can be implemented via a **try** clause followed by one or more **exception** clauses. They work similar to **if** and **else** clauses. Instead of evaluating a Boolean expression, Python checks if any of the exceptions specified in the `exception` clauses occur when the block in the `try` clause is executed. If so, some measures are taken to handle the error. Otherwise, program execution continues without further ado.

In our function for the trapezoidal rule, we can simply add `ZeroDivision Error` as an exception and print an error message *without* interrupting execution (explanation of function interface is omitted here):

```
1  def integr_trapez(f, a, b, n):
2
3      # integration step with exception handling
4      try:
5          h = (b - a)/n
6      except ZeroDivisionError:
7          print("Error: n must be non-zero")
8          return None
9
10     # endpoints of subintervals between a+h and b-h
11     x = np.linspace(a+h, b-h, n-1)
12
13     return 0.5*h*(f(a) + 2*np.sum(f(x)) + f(b))
14
15 print("  n   integr")
16 for n in range(0,60,10):
17     print("{:2d}".format(n), end="   ")
18
19     intgr = integr_trapez(np.sin, 0, np.pi/2; n)
20
21     if intgr != None:
22         print("{:.6f}".format(intgr))
```

Now we get the same output as above with an additional line indicating that zero subdivisons is not an allowed:

```
 n   integr
 0   Error: n must be non-zero
10   0.997943
20   0.999486
30   0.999772
40   0.999871
50   0.999918
```

This is accomplished by calculating the subinterval width in the `try` clause in lines 4–5. If n is zero, a `ZeroDivisionError` is encountered as exception (lines 6–8). After printing an error message, the function immediately returns None, indicating that the function did not arrive at a meaningful result for the actual arguments of

the function call. To place the error message in a single line right after the value of
n in the table, we split the print statement. First, the number of integration steps is
printed with end=" " in line 17. This replaces the default newline character by
two whitespaces, which separate the table columns. Then the value of the integral is
printed only if the call of integr_trapez() in line 19 returns a value that is not
None. Otherwise, the error message will appear.

We can even get more sophisticated though. In fact, only positive integers are
allowed for the number of subintervals. In a language such as C a variable of type
unsigned integer could be used and checking for zero values would be all that is
needed. Since a function argument in Python does not have a particular data type,
we need to convert n to an integer (assuming that the actual argument is at least a
number) and check that the result is positive.

```python
def integr_trapez(f, a, b, n):
    n = int(n)

    # integration step with exception handling
    try:
        if n > 0:
            h = (b - a)/n
        else:
            raise ValueError
    except ValueError:
        print("Invalid argument: n must be positive")
        return None

    # endpoints of subintervals between a+h and b-h
    x = np.linspace(a+h, b-h, n-1)

    return 0.5*h*(f(a) + 2*np.sum(f(x)) + f(b))
```

After chopping off any non-integer fraction with the help of int() in line 2, the
integration interval is subdivided in the try clause provided that n is greater than
zero (lines 6–7). If not, a ValueError is raised as exception. The programmer can
raise a specific exception via the keyword raise. We leave it as a little exercise for
you to test the function and to see what happens for arbitrary values of the argument
n. This kind of error checking might appear somewhat excessive for such a simple
application, but it can save you a lot of trouble in complex programs performing
lengthy computations. Making use of exception handling is considered to be good
programming practice and important for code robustness.

For Simpson's rule (3.20), we need to ensure that the number of subintervals, N, is
an even integer ≥ 2. The following implementation of Simpson's rule simply converts
any numeric value of the argument n into a number that fulfils the requirements of
Simpson's rule. Since the user of the function might not be aware of the assumptions
made about the argument, implicit changes of arguments should be used with care.
In contrast, the rejection of invalid arguments via exception handling usually tells
the user what the problem is. The downside is that exception handling takes more

effort and results in longer code. Here, we just make you aware of different options. You need to decide which methods is preferable depending on the purpose and target group of the code you write.

```
1   def integr_simpson(f, a, b, n):
2       """
3       numerical integration of a function f(x)
4       using Simpson's rule
5
6       args: f - function f(x)
7             a - left endpoint of interval
8             b - right endpoint of interval
9             n - number of subintervals (positive even integer)
10
11      returns: approximate integral
12      """
13
14      # need even number of subintervals
15      n = max(2, 2*int(n/2))
16
17      # integration step
18      h = (b - a)/n
19
20      # endpoints of subintervals (even and odd multiples of h)
21      x_even = np.linspace(a+2*h, b-2*h, int(n/2)-1)
22      x_odd  = np.linspace(a+h, b-h, int(n/2))
23
24      return (h/3)*(f(a) + 2*np.sum(f(x_even)) +
25                          4*np.sum(f(x_odd)) + f(b))
```

In line 15, any numeric value of n is converted into an even integer with a floor of two (test different values and figure out step by step how the conversion works). The expressions in lines 24–25 correspond to the terms in Eq. (3.20), with the elements of x_even being the endpoints x_{2k} with even indices (excluding x_0 and x_N), and x_odd containing those with odd indices, x_{2k-1}.

A test of integr_simpson() shows that high accuracy is reached with relatively few integration steps:

```
26  print(" n  integr")
27  for n in range(2,12,2):
28      print("{:2d}  {:.8f}".format(n,
29          integr_simpson(np.sin, 0, np.pi/2, n)))
```

```
 n  integr
 2  1.00227988
 4  1.00013458
 6  1.00002631
 8  1.00000830
10  1.00000339
```

The error of Simpson's rule is the order 10^{-5} for $N = 8$ compared to $N = 50$ for the trapezoidal rule. Consequently, the slightly more complicated algorithm pays off in terms of accuracy.

We are almost prepared now to solve the integral in Eq. (3.17) numerically. To apply numerical integration, we will use a slightly modified Python function for the Planck spectrum because the factor $1/\lambda^5$ and the exponent $hc/(\lambda kT)$ diverge toward the lower limit $\lambda = 0$ of the integral. However, analytically it follows that the combined factors do not diverge (the exponential function wins against the power function):

$$\lim_{\lambda \to 0} B_\lambda(T) = 0 \, .$$

Nevertheless, if you call `planck_spectrum()` defined in Sect. 3.1.2 for zero wavelength, you will encounter zero division errors. This can be avoided by shifting the lower limit of the integral from zero to a wavelength slightly above zero, for example, $\lambda = 1$ nm. Even so, Python will likely report a problem (it is left as an exercise to check this):

```
RuntimeWarning: overflow encountered in exp
```

The reason is that $hc/(\lambda kT)$ is a few times 10^3 for $\lambda = 1$ nm and typical stellar temperatures. For such exponents, the exponential is beyond the maximum that can be represented as a floating point number in Python (try, for instance, `np.exp(1e3)`).

We can make use of the `sys` module to obtain information about the largest possible floating point number:

```
30  import sys
31
32  print(sys.float_info.max)
```

It turns out to be a very large number:

```
1.7976931348623157e+308
```

Since the exponential function increases rapidly with the exponent, we need to ensure that the argument `np.exp()` does not exceed the logarithm of `sys.float_info.max`, which is just a little above 700 (assuming the value printed above). For this reason, we use the following definition of the Planck spectrum, where a cutoff of the exponent at 700 is introduced with the help of `np.minimum()`. This function compares its arguments element-wise and selects for each pair the smaller value:

```
33  def planck_spectrum(wavelength, T):
34      """
35      function computes Planck spectrum of a black body
36      uses cutoff of exponent to avoid overflow
37
38      args: numpy arrays
```

```
39              wavelength - wavelength in m
40              T - temperature in K
41
42          returns: intensity in W/m^2/m/sr
43          """
44          return 2*h*c**2 / (wavelength**5 *
45              (np.exp(np.minimum(700, h*c/(wavelength*k*T))) - 1))
```

With this modification, the Planck spectrum can be readily integrated in the interval $[\lambda_{min}, \lambda_0]$, where $\lambda_{min} = 1$ nm and $\lambda_0 = 364.7$ nm marks the Balmer jump. There is still a problem with applying our implementation of Simpson's rule, though. You can easily convince yourself that executing

```
integr_simpson(planck_spectrum, 1e-9, 364.7e-9, 100)
```

results in an error. To understand the problem you need to recall that `planck_spectrum()` is called in place of the generic function `f()` in the body of `integr_simpson()`. As a result, `planck_spectrum()` will be called with only one argument (the integration variable), but it expects the temperature as second argument. To solve this problem, we need to define a proxy function for `planck_spectrum()` which accepts the wavelength as single argument. This can be easily done by means of a Python lambda, which is also known as anonymous function. In essence, a Python lambda is a shorthand definition of a function, which can be directly used in an expression (so far, defining a function and calling the function in an expression have been distinct):

```
47  print("Teff [K] flux [%]")
48  for Teff in T_sample:
49      frac = np.pi*integr_simpson(
50          lambda x : planck_spectrum(x, Teff),
51          1e-9, 364.7e-9, 100) / (sigma * Teff**4)
52      print("{:5.0f}     {:5.2f}".format(Teff, 100*frac))
```

Instead of a function name, we use the expression beginning with the keyword `lambda` in line 50 as actual argument in the call of `integr_simpson()`. It defines the Planck spectrum for a given, but fixed temperature as an anonymous function, whose formal argument is `x`. For each temperature in `T_sample` (see end of Sect. 3.1.2), the result of the numerical intergration is multiplied by the factor $\pi/\sigma T_{eff}^4$ to obtain the fraction $F_{\lambda \leq \lambda_0}/F$:

```
Teff [K]  flux [%]
   3130     0.13
   3590     0.46
   4290     1.71
   5778     8.43
   9940    40.87
  24800    89.12
```

The percentages suggest that the Balmer jump should be prominent in the spectra of stars similar to our Sun and somewhat hotter. Although the fraction of ionizing photons increases with temperature, the amount of excited neutral hydrogen diminishes through thermal ionization at temperatures above 10000 K. This conclusion is in agreement with observed stellar spectra.

Exercises

3.3 Explain why the so-called Ca II H and K lines produced by singly ionized calcium in the ground state are so prominent in solar-type spectra, although the fraction of calcium atoms to hydrogen atoms is only 2×10^{-6}. The reasoning is similar to the analysis of Balmer lines in Sect. 3.2.1 (in fact, you will need results from this section to compare occupation numbers giving rise to K and H lines and Balmer lines). The ionization energy of neutral calcium (Ca I) is $\chi_1 = 6.11$ eV and the partition functions are $Z_I = 1.32$ and $Z_{II} = 2.30$. The energy difference between the ground state and the first excited state of Ca I is $E_2 - E_1 = 3.12$ eV with statistical weights $g_1 = 2$ and $g_2 = 4$.

3.4 Photons do not move straight through stellar interiors. They move only short distances before getting absorbed by atoms or ions, which subsequently re-emit photons. We can therefore think of a photon being frequently scattered in a stellar atmosphere. This process can be described by a simple model. If we assume that the photon moves in random direction over a constant length between two scattering events, its path can be described by a so-called random walk (see also [4], Sect. 9.3):

$$d = \ell_1 + \ell_2 + \cdots + \ell_N$$

where each step

$$\ell_n \equiv (\Delta x_n, \Delta y_n) = \ell \, (\cos \theta_n, \sin \theta_n)$$

has length ℓ and a random angle $\theta_n \in [0, 2\pi]$ (here, we consider only the two-dimensional case to simplify the model even further). After N steps, the expectation value for the net distance over which the photon has moved is $d = \ell \sqrt{N}$. Physically speaking, this is a diffusion process.

By making use of np.random.random_sample(size=N), you can generate an array of equally distributed random numbers in the interval [0, 1] of size N. By multiplying these numbers with 2π, you obtain a sequence of random angles for which you can compute a random walk according to the above formula. Set $\ell = 1/(\rho\kappa)$, where ρ is the density and κ the opacity of the gas. This is the mean free path. The density and opacity in the layers beneath the atmosphere of a solar-mass star is roughly $\rho = 10^{-6}$ g cm^{-3} and $\kappa = 50$ cm^2 g^{-1}.

(a) Compute a random walk and use plt.plot() with 'o-' as marker to show all positions (x_n, y_n) as dots connected by lines.
(b) Compute a series of random walks with N increasing in logarithmic steps and determine the distance $d = |d|$ between start end points for each walk. The function

```
curve_fit(func, xdata, ydata)
```

from `scipy.optimize` allows you to fit data points given by `xdata` and `ydata`, respectively, to a model that is defined by the Python function `func()`. Arguments of the function are an independent variable, which corresponds to `xdata`, and one or more free parameters. Fitting data to a model means that the parameters with the smallest deviation between data and model are determined, which is an optimization problem (we do not worry here about the exact mathematical meaning of deviation).[21] To apply `curve_fit()`, you need to collect the data for N and d from your random walks in arrays, and define a Python function for the expectation value $d = \ell\sqrt{N}$ with parameter ℓ. How does the resulting value of ℓ compare to the mean-free path calculated above? That is to say if you knew only the random-walk data, would you be able to estimate the mean-free path from the fit? Plot your data as dots and the resulting fit function as curve to compare data and model.

(c) The total travel time of the photon over a random walk with N steps is $t = N\ell c$, where c is the speed of light. How long would it take a photon to reach the photosphere, from which radiation is emitted into space, from a depth of 10^4 km (about 1% of the solar radius)?

3.3 Planetary Ephemerides

The trajectory followed by an astronomical object is also called ephemeris, which derives from the Latin word for "diary". In former times, astronomers observed planetary positions on a daily basis and noted the positions in tables. If you think about the modern numerical computation of an orbit, which will be discussed in some detail in the next chapter, you can imagine positions of a planet or star being chronicled for subsequent instants, just like notes in a diary.

In Sect. 2.2, we considered the Keplerian motion of a planet around a star, which neglects the gravity of other planets. The orbits of the planets in the solar system can be treated reasonably well as Kepler ellipses, but there are long-term variations in their shape and orientation induced by the gravitational attraction of the other planets, especially Jupiter. To compute such secular variations over many orbital periods with high accuracy, analytical and numerical calculations using perturbation techniques are applied. An example is the VSOP (Variations Séculaires des Orbites Planétaires) theory [8]. The solution known as VSOP87 represents the time-dependent heliocentric coordinates X, Y, and Z of the planets in the solar system (including Pluto) by series expansions, which are available as source code in several programming languages. We produced a NumPy-compatible transcription into Python.

The Python functions for the VSOP87 ephemerides are so lengthy that it would be awkward to copy and paste them into a notebook. A much more convenient option

---

[21] It might be helpful to study the explanation and examples in
docs.scipy.org/doc/scipy/reference/generated/scipy.optimize.curve_fit.html.

is to put the code into a user defined Python module. In its simplest manifestation, a module is just a file named after the module with the extension .py containing a collection of function definitions. The file vsop87.py is part of the zip archive accompanying this chapter. You can open it in any source code editor or IDE. After a header, you will find definitions of various functions and, as you scroll down, thousands of lines summing cosine functions of different arguments (since we use the cosine from numpy, we need to import this module in vsop87.py). As with other modules, all you need to do to use the functions in a Python script or in a notebook is to import the module:

```
1  import vsop87 as vsop
```

However, this will only work if the file vsop87.py is located in the same directory as the script or notebook into which it is imported. If this is not the case, you can add the directory containing the file to the module search path. Python searches modules in predefined directories listed in sys.path You can easily check which directories are included by importing the sys module and printing sys.path. If you want to add a new directory, you need to append it to the environmental variable PYTHONPATH before starting your Python session. The syntax depends on your operating system and the shell you are using (search the web for pythonpath followed by the name of your operating system and you are likely to find instructions or some tutorial explaining how to proceed). Alternatively, you can always copy vsop87.py into the directory you are currently working.

Functions such as vsop.Earth_X(), vsop.Earth_Y(), vsop.Earth_Z() for the coordinates of Earth expect the so-called Julian date as argument. The Julian date is used by astronomers as a measure of time in units of days counted from a zero point that dates back to the fifth millennium B.C. (as a consequence, Julian dates for the current epoch are rather large numbers). The following Python function converts the commonly used date of the Gregorian calendar (days, months, year) and universal time (UT) to the corresponding Julian date:

```
2  import math
3
4  def Julian_date(D, M, Y, UT):
5      """
6      converts day, month, year, and universal time into
7      Julian date
8
9      args: D - day
10           M - month
11           Y - year
12           UT - universal time
13
14     returns: Julian date
15     """
16
17     if (M <= 2):
18         y = Y-1
19         m = M+12
```

```
20        else:
21            y = Y
22            m = M
23
24        if (Y < 1582):
25            B = -2
26        elif (Y == 1582):
27            if (M < 10):
28                B = -2
29            elif (M == 10):
30                if (D <= 4):
31                    B=-2
32                else:
33                    B = math.floor(y/400) - math.floor(y/100)
34            else:
35                B = math.floor(y/400) - math.floor(y/100)
36        else:
37            B = math.floor(y/400) - math.floor(y/100)
38
39        return math.floor(365.25*y) + math.floor(30.6001*(m+1)) + \
40            B + 1720996.5 + D + UT/24
```

The function `math.floor()` returns the largest integer less than or equal to a given floating point number.[22]

For example, let us the determine the current Julian date:

```
41  from datetime import datetime
42
43  # get date and UTC now
44  now = datetime.utcnow()
45
46  JD = Julian_date(now.day, now.month, now.year,
47                   now.hour + now.minute/60 + now.second/3600)
48
49  # convert to Julian date
50  print("Julian date: {:.4f}".format(JD))
```

The function `datetime.utcnow()` returns the coordinated universal time (see Sect. 2.1.3) at the moment the function is called. From the object `now` defined in line 44, we can get the calendar day via the attribute `day`, the hour of the day via `hour`, etc. This is the input we need for converting from UTC to the Julian date. Since the argument `UT` of `Julian_date()` must be the time in decimal representation, we need to add the number of minutes divided by 60 and the number of seconds divided by 3600. When this sentence was written, the result was

```
Julian date: 2458836.0753
```

Alternatively, you can use Astropy (see Exercise 3.6).

---

[22]This is is not identical to `int()`, which chops off the non-integer fraction. Apply both functions to some negative floating point number to see the difference.

For a given Julian date, we can easily compute the distance between two planets by using the VSOP87 coordinate functions of the planets. VSOP87 uses an ecliptic heliocentric coordinate system with the Sun at the center and the ecliptic being coplanar with the XY plane (i.e. Earth's orbit is in the XY plane). For example, the distance between Earth ($\oplus$) and Mars (σ) is given by

$$d = \sqrt{(X_\oplus - X_\sigma)^2 + (Y_\oplus - Y_\oplus)^2 + (Z_\oplus - Z_\sigma)^2}$$

which translates into the following Python code:

```
51  def Earth_Mars_dist(JD):
52      delta_x = vsop.Earth_X(JD)  -  vsop.Mars_X(JD)
53      delta_y = vsop.Earth_Y(JD)  -  vsop.Mars_Y(JD)
54      delta_z = vsop.Earth_Z(JD)  -  vsop.Mars_Z(JD)
55      return vsop.np.sqrt(delta_x**2 + delta_y**2 + delta_z**2)
```

At first glance, you might find it surprising that NumPy's square root is called as `vsop.np.sqrt()` in the last line. But remember that `numpy` is imported (under the alias np) inside the module `vsop`, while the function `Earth_Mars_dist()` is is not part of this module.[23] For this reason, NumPy functions such as `sqrt()` need to be referenced via `vsop.np` (dotted module names are also used in Python packages consisting of a hierarchy of modules). Of course, we could import `numpy` directly into the global namespace. In this case, each NumPy function would have a duplicate in `vsop.np` (check that both variants work).

Now execute `Earth_Mars_dist()` for your Julian date and print the result:

```
56  print("distance between Earth and Mars now: {:.3f} AU".\
57          format(Earth_Mars_dist(JD)))
```

The answer for the Julian date 2458836.0753 is

```
distance between Earth and Mars now: 2.278 AU
```

Since VSOP87 computes the coordinates in AU, no unit conversion is required.
A plot showing d for the next 1000 days is easily produced:

```
58  import matplotlib.pyplot as plt
59  %matplotlib inline
60
61  t = JD + np.arange(1000)
62
63  plt.figure(figsize=(6,4), dpi=100)
64  plt.plot(t, Earth_Mars_dist(t))
65  plt.xlabel("JD [d]")
66  plt.ylabel("$d$ [AU]" )
```

---

[23]It is possible to add a new function to the module by inserting it into the file `vsop87.py` and reloading the module.

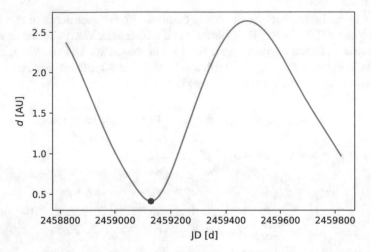

Fig. 3.5 Distance between Earth and Mars over 1000 days. Time is expressed as Julian date. The next minimum is indicated by the red dot

Since the Julian data counts days, it is sufficient to add days counting from 0 to 999 to the current date JD (see line 61). By applying `np.arange()`, we obtain an array of dates for which distances can be directly computed and plotted. The resulting graph will look similar to what is shown in Fig. 3.5 (it will be shifted depending on the chosen start date). The distance between the two planets varies with their orbital motion and becomes minimal when Earth is located just in between Mars and Sun (Mars and Sun are then said to be in opposition as seen from Earth). To determine the date when this happens, we need to find points where the first derivative with respect to time, $\dot{d}$, vanishes. This brings us back to root finding.

If you look into `vsop87.py`, you will realize that it would be unpractical to calculate derivatives of the coordinate functions analytically. However, we can approximate the derivatives numerically by finite differences. We concentrate on centered differences, which are obtained by averaging the forward and backward differences of a function $f(x)$ for discrete points on the x-axis spaced by Δx. To second order in Δx,[24] the derivative is approximated by the centered difference quotient

$$f'(x) \equiv \frac{\mathrm{d}f}{\mathrm{d}x} \simeq \frac{f(x + \Delta x) - f(x - \Delta x)}{2\Delta x}. \tag{3.21}$$

The following Python function implements the centered difference method for a single point or an array of points (the backward and forward coordinate difference $h = \Delta x$ is specified as third argument).

---

[24]This means that the error decreases with Δx^2 as $\Delta x \to 0$, provided that the function is sufficiently smooth to be differentiable.

```
1  def derv_center2(f, x, h):
2      """
3      approximates derivative of a function
4      by second-order centered differences
5
6      args: f - function f(x)
7            x - points for which df/dx is computed
8            h - backward/forward difference
9
10     returns: approximation of df/dx
11     """
12     return (f(x+h) - f(x-h))/(2*h)
```

As a simple test, we compute the derivative of the sine function. Since the derivative of $\sin(x)$ is $\cos(x)$, we can compare the numerical approximation to the analytic solution:

```
13 h = 0.1
14 x = np.linspace(0, np.pi, 9)
15
16 print(" analytic    cd2")
17 for (exact,approx) in zip(np.cos(x),
18                           derv_center2(np.sin, x, h)):
19     print("{:9.6f} {:9.6f}".format(exact,approx))
```

By passing the NumPy array x defined in line 14 as argument, np.cos() and derv_center2() return arrays of analytic and approximate values, respectively, which are zipped and printed in a table via a **for** loop. The label cd2 is an abbreviation for centered differences of second order. In this case, the variable h is not given by the spacing of the points in the array x (the example above, the spacing is $\pi/8$). Its value is used as an adjustable parameter to control the accuracy of the centered difference approximation.[25] For $h = 0.1$, the results are:

```
    analytic    cd2
    1.000000   0.998334
    0.923880   0.922341
    0.707107   0.705929
    0.382683   0.382046
    0.000000   0.000000
   -0.382683  -0.382046
   -0.707107  -0.705929
   -0.923880  -0.922341
   -1.000000  -0.998334
```

---

[25] In many applications, however, h is equal to the spacing of grid points. This typically occurs when a function cannot be evaluated for arbitrary x-values, but is given by discrete data.

You can gradually decrease the value of h to investigate how the finite difference approximation converges to the derivative (remember that the derivative is defined by the differential quotient in the limit $h \to 0$).

In order to determine whether the function $f(x)$ has a minimum or a maximum, we need to evaluate the second derivative, $f''(x)$. The second-order centered difference approximation,

$$f''(x) \simeq \frac{f(x + \Delta x) - 2f(x) + f(x - \Delta x)}{\Delta x^2} , \qquad (3.22)$$

is also readily implemented in Python:

```
20  def dderv_center2(f, x, h):
21      """
22      approximates second derivative of a function
23      by second-order centered differences
24
25      args: f - function f(x)
26            x - points for which df/dx is computed
27            h - backward/forward difference
28
29      returns: approximation of d^2 f/dx^2
30      """
31      return (f(x+h) - 2*f(x) + f(x-h))/h**2
```

With the help of dderv_center2(), you can elaborate on the example above and determine points for which the sine function has a minimum, a maximum, or an inflection point. This is left as an exercise for you (convince yourself that your numerical results are consistent with the extrema and inflection points following from analytic considerations).

Returning to our problem of finding the next date when Mars is closest to Earth, the numerical computation of the second derivative comes in very handy if we apply yet another method to find the root of a function. This method, which is known as Newton–Raphson method (or Newton's method), makes use of tangent lines to extrapolate from some given point to the point where the function crosses zero. If $f(x)$ is a linear function and $f'(x)$ its constant slope, it is a matter of elementary geometry to show that $f(x_1) = 0$ for

$$x_1 = x - \frac{f(x)}{f'(x)} . \qquad (3.23)$$

Of course, we are interested in functions that are non-linear and for which $f(x) = 0$ cannot be found analytically. Assuming that $f(x)$ has a root in some neighbourhood of the point x_0 and provided that $f(x)$ is differentiable in this neighbourhood, the formula for a linear function can be applied iteratively:

$$x_{n+1} = x_n - \frac{f(x_n)}{f'(x_n)} . \tag{3.24}$$

As the root is approached, $f(x_n)$ converges to zero and the difference between x_n and x_{n+1} vanishes. If a function has multiple roots, the result depends on the choice of the start point x_0.

In Python, the algorithm can be implemented as follows:

```python
def root_newton(f, df, x0, eps=1e-3, imax=100):
    """
    Newton-Raphson algorithm for finding the root
    of a function f(x)

    args: f - function f(x)
          df - derivative df/dx
          x0 - start point of iteration
          eps - tolerance
          imax - maximal number of iterations
          verbose - print additional information if true

    returns: estimate of x for which f(x) = 0
    """

    for i in range(imax):
        x = x0 - f(x0)/df(x0)

        if abs(x - x0) < eps:
            print("tolerance reached after {:d} iterations".
                    format(i+1))
            print("deviation: f(x) = {:.3e}".format(f(x)))
            return x

        x0 = x

    print("exceeded {:d} iterations".format(i+1),
          "without reaching tolerance")
    return x
```

The function defined above is similar to the first version of `root_bisection()` in Sect. 3.1.2. Instead of the endpoints of the start interval for the bisection method, a single start point `x0` has to be specified, and in addition to the function `f()` we need to pass its derivative `df()` as argument. The body of `root_newton()` is quite simple: The iteration formula (3.24) is repeatedly applied in a **for** loop with a prescribed maximum number of iterations (optional argument `imax`). Once the difference between the previous estimate `x0` and the current estimate `x` is smaller than the tolerance `eps`, the execution of the loop stops and `x` is returned by the function. Otherwise, the function terminates with an error message.

To test the Newton–Raphson method, let us return to the quadratic function

```
61  def quadratic(x):
62      return x**2 - x - 2
```

which we used as test case for the bisection method. Choosing the start point $x_0 = 0$,

```
63  root_newton(quadratic, lambda x: 2*x - 1, 0)
```

we get the first root as solution:

```
tolerance reached after 5 iterations
deviation: f(x) = 2.095e-09


-1.000000000698492
```

Here, the derivative

$$f'(x) = 2x - 1$$

is defined in the call of Newton's method via a Python lambda (see Sect. 3.2.1; alternvatively, you can define the derivative separately as a named Python function). For $x_0 = 10$, on the other hand,

```
64  root_newton(quadratic, lambda x: 2*x - 1, 10)
```

we get

```
tolerance reached after 5 iterations
deviation: f(x) = 1.267e-08


2.0000000042242396
```

Compared to the bisection method, Newton's method produces highly accurate approximations after only a few iterations. The downside is that one cannot predict which root is obtained for a given start point and a recursive variant for multiple roots as in the case of the bisection method is not possible.

In the example above, the function and its derivative are defined by analytic expressions. For the distance between Earth and Mars, we use centered differences to compute derivatives numerically. By applying the Newton–Raphson method to the first time derivative of the distance, i.e. $\dot{d}(t)$ in place of $f(x)$, we get an extremum of $d(t)$. Instead of $f'(x)$, the second derivative of the distance $\ddot{d}(t)$ is required as input. To calculate the derivatives, we apply `derv_center2()` and `dderv_center2()`:

```
65  delta_t = 0.1
66
67  JD_extrem = root_newton(
68      lambda t : derv_center2(Earth_Mars_dist, t, delta_t),
69      lambda t : dderv_center2(Earth_Mars_dist, t, delta_t),
70      JD+300, eps=delta_t)
71
```

```
72  print("\ndistance = {1:.3f} AU in {0:.0f} days".\
73         format(JD_extrem-JD, Earth_Mars_dist(JD_extrem)),
74         "({:4.0f}-{:02.0f}-{:02.0f})".\
75         format(vsop.JD_year(JD_extrem),
76                vsop.JD_month(JD_extrem),
77                vsop.JD_day(JD_extrem)))
```

To understand the call of `root_newton()`, it is important to keep in mind that functions with only a single argument are assumed in line 48 in the body of `root_newton()`. However, `derv_center2()` and `dderv_center2()` have several arguments (see definitions above). For this reason, we use `lambda` to define anonymous functions of the variable `t` returning centered difference approximations for a given function (`Earth_Mars_dist`) and timestep (`delta_t`). The Newton–Raphson iteration starts at the current Julian date `JD` plus 300 days (see line 70). In our example, this results in

```
tolerance reached after 3 iterations
deviation: f(x) = 6.209e-13


distance = 0.415 AU in 307 days (2020-10-06)
```

It turns out that the initial guess of 300 days was pretty close. To print the corresponding date in standard format, the Julian date is converted with the help of `JD_year`, `JD_month`, and `JD_day` from `vsop87` (lines 74–77). The solution is shown as red dot in Fig. 3.5 (we leave it to you to complete the code). It turns out that the distance to Mars reaches a minimum in October 2020.

The solution you get will depend on your initial date and start point for the Newton–Raphson method. To verify whether your result is a minimum or a maximum, evaluate

`dderv_center2(Earth_Mars_dist, JD_extrem, delta_t)`

If the sign of the second derivative is positive, the distance is minimal, otherwise it is maximal. At the time of its closet approach, Mars appears particularly bright on the night sky. If you are enthusiastic about observing the sky, the date following from your calculation might be worthwhile to note.

Exercises

3.5 Apply centered differences to calculate the fastest change of the day length in minutes per day at your geographical latitude (see Sect. 2.1.2).

3.6 Planetary ephemerides are included in `astropy.coordinates`. For example, to get the barycentric coordinates of Mars at a particular date,[26] you can use the code

---

[26]The origin of barycentric coordinates is the center of mass of the solar system. This coordinate frame is also known as International Celestial Reference System (ICRS).

```
from astropy.time import Time
from astropy.coordinates import solar_system_ephemeris, \
    get_body_barycentric

solar_system_ephemeris.set('builtin')

get_body_barycentric('mars', Time("2019-12-18"))
```

In this case, the computation is based on Astropy's built-in ephemeris.[27]

Rewrite Earth_Mars_dist() using get_body_barycentric() for the computation of the distance from Earth to Mars and produce a plot similar to Fig. 3.5. For the time axis, you can define an array of days similar to elapsed in Sect. 2.1.3. When calling Earth_Mars_dist(), you need to add the array to a Time object. You can express time also as Julian date by using the **format** attribute.[28]

3.7 Starting from an initial date of your choice, investigate possible Hohmann transfer orbits to Mars (see Exercise 2.10) over a full orbital period (687 days). For each day, use trigonometry and the ephemerides of Earth and Mars to compute the angular separations δ between the two planets at launch time and the angular change $\Delta\varphi = \varphi' - \varphi$ in the position of Mars over the transfer time t_H. Neglecting the inclination of the orbital plane of Mars against the ecliptic, these two angles should add up to 180° (one half of a revolution) for a Hohmann transfer, i.e. $\delta + \Delta\varphi = 180°$. Which launch dates come closest to satisfying this condition? However, the parameters of the transfer trajectory are based on the assumption that the orbit of Mars is circular. In fact, Mars has the highest eccentricity of all planets in the solar system. Based on radial distances from the Sun, estimate by what distance is the spaceship going to miss Mars when it reaches the aphelion.[29]

3.8 From all planets in the solar system, what is the largest distance between two planets going to be in the course of the next 165 years (the time needed by Neptune to complete one full revolution around the Sun)?

---

[27] There are options for more precise positions; see
docs.astropy.org/en/stable/coordinates/solarsystem.html for further details.

[28] See docs.astropy.org/en/stable/time.

[29] Determining a trajectory with sufficient accuracy for a space mission is quite challenging and requires exact orbital elements and solutions of the equations of motion. Moreover, it is a complicated optimization problem to meet constraints such as fuel consumption and travel time. As a result, various modifications are made. For example, other types of Hohman transfer trajectories intersect the orbit of the target prior to or after reaching the aphelion. Many space missions, such as the famous Voyager missions, utilize the gravity of planets during flybys to alter their trajectories.

Chapter 4
Solving Differential Equations

Abstract Differential equations play a central role in numerics. We introduce basic algorithms for first and second-order initial value problems. This will allow us to numerically solve many interesting problems, for instance, a body falling through the atmosphere, two- and three-body problems, a simple model for galaxy collisions, and the expansion of the Universe. Apart from learning how to deal with numerical errors, mastering multi-dimensional arrays and complex operations are important objectives of this chapter. Moreover, it is shown how to produce histograms and three-dimensional plots.

4.1 Numerical Integration of Initial Value Problems

From Newton's laws to Schrödinger's equation: physics is packed with differential equations. While the majority are partial differential equations, such as Euler's equations of fluid dynamics, Maxwell's equations for electromagnetic fields, and Schrödinger's equation for the wave function in quantum physics, there also are many applications of ordinary differential equations. In contrast to partial differential equations, ordinary differential equations determine functions of a single variable. In this section, you will learn how to solve such equations numerically.

4.1.1 First Order Differential Equations

A differential equation of first order determines a time-dependent function $x(t)$ by a relation between the function and its first derivative $\dot{x}$. An example is the equation for radioactive decay:

$$\dot{x} = \lambda x , \qquad (4.1)$$

© Springer Nature Switzerland AG 2021
W. Schmidt and M. Völschow, *Numerical Python in Astronomy and Astrophysics*,
Undergraduate Lecture Notes in Physics,
https://doi.org/10.1007/978-3-030-70347-9_4

The solution of this equation is $x(t) = x_0 e^{-\lambda t}$, where $x_0 = x(0)$ is said to be the initial value. While it is straightforward to solve a linear differential equation, in which all terms are linear in x or $\dot{x}$, non-linear differential equations are more challenging.

Consider the Bernoulli equation[1]:

$$\dot{x} = \alpha(t)x + \beta(t)x^\rho \,, \tag{4.2}$$

where $\alpha(t)$ and $\beta(t)$ are given functions and ρ is a real number. It encompasses important differential equations as special cases, for example, Eq. (4.1) follows for $\alpha(t) = \lambda$, $\beta(t) = 0$, $\rho = 0$ and the logistic equation describing population dynamics for $\alpha(t) = a$, $\beta(t) = b$ with constants $a > 0$, $b < 0$, and $\rho = 2$ (see Exercise 4.1). From the viewpoint of numerics, Bernoulli-type equations are interesting because an analytic solution is known and can be compared to numerical approximations.

In the following, we will attempt to numerically solve an example for a Bernoulli equation from astrophysics, namely the equation for the radial expansion of a so-called Strömgren sphere. When a hot, massive star is borne, it floods its surroundings with strongly ionizing UV radiation. As a result, a spherical bubble of ionized hydrogen (H II) forms around the star. It turns out that ionization progresses as ionization front, i.e. a thin spherical shell propagating outwards (see [4], Sect. 12.3). Inside the shell, virtually all hydrogen is ionized. The radial propagation of the shell is described by a differential equation for the time-dependent radius $r(t)$ [9] (convince yourself that this equation is a Bernoulli differential equation):

$$\dot{r} = \frac{1}{4\pi r^2 n_0}\left(S_* - \frac{4\pi}{3}r^3 n_0^2 \alpha\right) \tag{4.3}$$

Here, S_* is the total number of ionizing photons (i.e. photons of energy greater than 13.6 eV; see Sect. 3.2.1) per unit time, n_0 is the number density of neutral hydrogen atoms (H I), and $\alpha \approx 3.1 \times 10^{-13}$ cm^3 s^{-1} is the recombination coefficient. Recombination of ionized hydrogen and electrons competes with ionization of neutral hydrogen.[2]

Equation (4.3) can be rewritten in the form

$$\dot{r} = n_0 \alpha \,\frac{r_s^3 - r^3}{3r^2} \tag{4.4}$$

where

$$r_s = \left(\frac{3S_*}{4\pi n_0^2 \alpha}\right)^{1/3} \tag{4.5}$$

---

[1]Named after the Swiss mathematician Jakob Bernoulli, one of the pioneers of calculus in the late 17th century.

[2]The recombination rate is proportional to the product of the number densities of hydrogen ions and electrons, which is n_0^2 in the case of a fully ionized medium. The total number of recombinations per unit time is obtained by multiplying with the volume of the sphere.

is the Strömgren radius. The speed $\dot{r}$ approaches zero for $r \rightarrow r_s$, i.e. the propagation of the ionization front slows down and stalls at the Strömgren radius.[3] How large is a Strömgren sphere? Let us do the calculation for an O6 star (see also [4], example 12.4):

```
1  import numpy
2  import astropy.units as unit
3
4  n0 = 5000 * 1/unit.cm**3 # number density of HI
5  S = 1.6e49 * 1/unit.s # ionizing photons emitted per second
6  alpha = 3.1e-13 * unit.cm**3/unit.s # recombination coefficient
7
8  rs = (3*S/(4*np.pi * n0**2 * alpha))**(1/3)
9  print("Strmoegren radius = {:.2f}".format(rs.to(unit.pc)))
```

The photons emitted by the star per second can be estimated from the luminosity and the peak of the Planck spectrum for an effective temperature $T_{eff} \approx 4.5 \times 10^4$ K. Astropy's units module allows us to express the result in parsec without bothering about conversion factors (see also Sect. 3.1.1):

```
Stroemgren radius = 0.26 pc
```

To integrate Eq. (4.4) in time, we can apply a linear approximation over a small time interval $[t, t + \Delta t]$ (also called time step):

$$r(t + \Delta t) \simeq r(t) + \dot{r}(t)\Delta t \tag{4.6}$$

In other words, it is assumed that the derivative $\dot{r}$ changes only little over Δt and can be approximated by the value at time t. For given $r(t)$, we can then substitute the right-hand side of the differential equation for $\dot{r}(t)$:

$$r(t + \Delta t) \simeq r(t) + n_0\alpha \frac{r_s^3 - r(t)^3}{3r(t)^2}\Delta t$$

Starting with an initial value $r(0) = r_0$, this rule can be applied iteratively for $t_n = n\Delta t$. This is the basic idea of the Euler method.[4] For a general first-order differential equation

$$\dot{x} = f(t, x), \tag{4.7}$$

it can be written as iterative scheme

$$x_{n+1} = x_n + f(t_n, x_n)\Delta t \tag{4.8}$$

---

[3] In fact, the sphere of ionized hydrogen begins to expand once $\dot{r}$ drops below the speed of sound. The expansion stops when it reaches pressure equilibrium with the surrounding neutral medium.

[4] The method was introduced by Leonhard Euler in his influential textbook on calculus from 1768, long before it was possible to routinely carry out numerical approximations with the help of computers.

where $x_n = x(t_n)$ and $n = 0, 1, 2, \ldots$

In the following code, we iteratively compute the function values r_n in a `for` loop and collect them in a NumPy array, which is useful for plotting:

```
10  n0_cgs = n0.value
11  alpha_cgs = alpha.value
12  rs_cgs = rs.value
13
14  # time step in s
15  dt = 100
16  n_steps = 1000
17
18  # intialization of arrays for t and r(t)
19  t = np.linspace(0, n_steps*dt, n_steps+1)
20  r = np.zeros(n_steps+1)
21
22  # start radius in cm
23  r[0] = 1e16
24
25  # Euler integration
26  for n in range(n_steps):
27      rdot = n0_cgs * alpha_cgs * \
28              (rs_cgs**3 - r[n]**3)/(3*r[n]**2)
29      r[n+1] = r[n] + rdot * dt
```

The loop begins with n=0 and the first array element r[0] defined in line 23. To compute the next value using the Euler scheme, the time derivative rdot is evaluated in lines 27–28. The new radius is then assigned to the next element of the array r. The array length is given by the number of time steps n_steps plus one element for the initial value (line 20). In lines 10–12, all parameters are converted to pure floating point numbers because Astropy units cannot be used with individual array elements (see Sect. 2.1.3). In contrast to Python lists, which support arbitrary elements, array elements must be floating point numbers without additional attributes. As demonstrated in Appendix B.1, there is a significant trade-off in terms of computational efficiency and memory consumption. In the example above, it is understood that all variables are in cgs units.

Since the right-hand side of Eq. (4.4) diverges for $r = 0$, the initial radius r_0 must be positive. A reasonable choice is a small fraction of r_s, which is of the order of a parsec (about 3×10^{18} cm). We set $r_0 = 10^{16}$ cm. Given this initial value, what is an appropriate choice for the time step Δt? As a first guess, we set the time step to 100 s (see line 15). The numerical solution for 1000 time steps is plotted with the following code as dashed line. The plot is shown in Fig. 4.1.

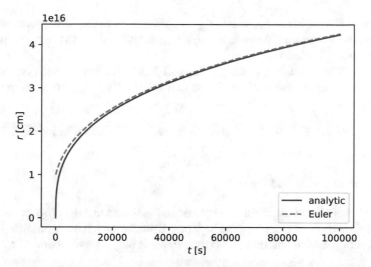

Fig. 4.1 Numerical solution of the initial value problem (4.4) for a Strömgren sphere with $r_0 = 10^{16}$ cm using the Euler method (dashed line). The analytic solution for $r_0 = 0$ is shown as solid line

```
30  import matplotlib.pyplot as plt
31
32  fig = plt.figure(figsize=(6, 4), dpi=100)
33  plt.plot(t,
34          rs_cgs*(1.0 - np.exp(-n0_cgs*alpha_cgs*(t)))**(1/3),
35          linestyle='-' , color='red' , label="analytic")
36  plt.plot(t, r, linestyle='--' , color='green' , label="Euler")
37  plt.legend(loc='lower right')
38  plt.xlabel("$t$ [s]")
39  plt.ylabel("$r$ [cm]")
40  plt.savefig("stroemgren_cgs.pdf")
```

For comparison, the analytic solution [9]

$$r(t) = r_{\mathrm{s}} \left(1 - \mathrm{e}^{-n_0 \alpha t}\right)^{1/3} \tag{4.9}$$

is plotted in lines 33–35 (see solid line in Fig. 4.1). It agrees quite well with our numerical solution, except for the discrepancy at early time. Actually, this is mainly caused by different zero points of the time coordinates. For numerical integration, we assume the initial value $r(0) = r_0$, while formula (4.9) implies $r(0) = 0$. This can be fixed by shifting the time coordinate of the analytic solution such that $r(0) = r_0$. Solving Eq. (4.9) with $r = r_0$ for time yields

$$t_0 = \frac{1}{n_0 \alpha} \log \left[1 - (r_0/r_{\mathrm{s}})^3\right] . \tag{4.10}$$

By plotting $r(t') = r(t - t_0)$, where $t' \geq 0$ corresponds to the time coordinate used for numerical integration, you will find that the analytical solution is closely matched by the numerical solution.

The coefficient $1/n_0\alpha$ appearing in Eq. (4.10) is a time scale, which can be interpreted as formation time t_s of the Strömgren sphere. By plugging in the numbers,

```
41  ts = 1/(n0*alpha)
42  print("Time scale = {:.2f}".format(ts.to(unit.yr)))
```

we find

```
Time scale = 20.44 yr
```

For a radius of about 0.3 pc, this is very fast and indicates that the propagation speed of the ionization front must be quite high (see Exercise 4.2). It turns out that the final time of our numerical solution (1000 time steps) is only a tiny fraction of t_s:

```
43  t[-1]*unit.s/ts
```

The output is

```
0.000155
```

As a result, we would need roughly $10^4 \times 1000 \sim 10^7$ time steps to reach t_s. This is a very large number of steps. To compute the time evolution for a longer interval of time, we can increase the time step Δt, which in turn reduces the number of time steps. However, keep in mind that Eq. (4.8) is an approximation. We need to weigh the reduction of computational cost (fewer steps) against the loss of accuracy (larger time step).

For a sensible choice, it is important to be aware of the physical scales characterizing the system. The differential equation for the time-dependent radius of a Strömgren sphere can be expressed in terms of the dimensionless variables $\tilde{r} = r/r_s$ and $\tilde{t} = t/t_s$:

$$\frac{d\tilde{r}}{d\tilde{t}} = \frac{1 - \tilde{r}^3}{3\tilde{r}^2} . \tag{4.11}$$

In this form, it is much easier to choose initial values and an appropriate integration interval. The initial radius should be small compared to the Strömgren radius, for example, $\tilde{r}_0 = 0.01$. To follow the evolution of the sphere over the time scale t_s, we need to integrate at least over the interval $[0, 1]$ with respect to $\tilde{t}$ (in the example above, the interval was $[0, 0.000155]$). Obviously, the times step must be small compared to t_s, i.e. $\Delta t/t_s \equiv \Delta\tilde{t} \ll 1$. The following code computes and plots the numerical solution for different time steps, starting with $\Delta\tilde{t} = 10^{-3}$.

```python
44  # initial radius (dimensionless)
45  r0 = 0.01
46
47  # analytic solution
48  t0 = np.log(1 - r0**3)
49  t = np.arange(0, 2.0, 0.01)
50
51  fig = plt.figure(figsize=(6, 4), dpi=100)
52  plt.plot(t, (1.0 - np.exp(-t+t0))**(1/3),
53           color='red', label="analytic")
54
55  # time step (dimensionless)
56  dt = 1e-3
57  n_steps = 2000
58
59  while dt >= 1e-5:
60      t = np.linspace(0, n_steps*dt, n_steps+1)
61      r = np.zeros(n_steps+1)
62      r[0] = r0
63
64      print("Integrating {:d} steps for dt = {:.0e}".
65            format(n_steps,dt))
66      for n in range(n_steps):
67          rdot = (1 - r[n]**3)/(3*r[n]**2)
68          r[n+1] = r[n] + rdot * dt
69
70      # plot the data
71      plt.plot(t, r, linestyle='--',
72               label="Euler, $\Delta t$ = {:.1f}".
73                     format(dt*ts.to(unit.hr)))
74
75      # decrease time step by a factor of 10
76      dt *= 0.1
77      n_steps *= 10
78
79  plt.legend(loc='lower right')
80  plt.xlabel("$L/L_{\mathrm{s}}$")
81  plt.ylabel("$r/r_{\mathrm{s}}$")
82  plt.ylim(0,1)
83  plt.savefig("stroemgren_dimensionless.pdf")
```

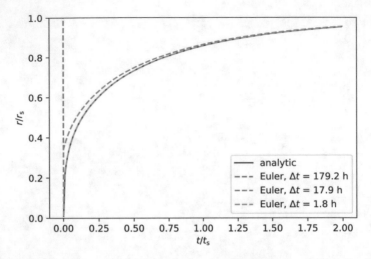

Fig. 4.2 Numerical solutions of the differential equation in dimensionless formulation (4.11) for different time steps ($\Delta \tilde{t} = 10^{-3}$, 10^{-4}, and 10^{-5}; the corresponding physical time steps are indicated in the legend). The Strömgren radius and formation time are $r_s = 0.26$ pc and $t_s = 20.4$ yr, respectively

This examples shows how to plot multiple graphs from within a loop. The time step is iterated in a `while` loop. The nested `for` loop starting at line 66 applies the Euler scheme to Eq. (4.11). Here, the arrays `t` and `r` contain values of the dimensionless variables $\tilde{t}$ and $\tilde{r}$. It is important to create these arrays for each iteration of the outer loop (see lines 60–61) because the number of time steps and, consequently, the array size increases. After plotting `r` versus `t`, the time step is reduced by a factor of 10 (line 76) and the number of steps increases by a factor of 10 to reach the same end point (line 77). The `while` loop terminates if $\Delta \tilde{t} < 10^{-5}$. The physical time step in hours is printed in the legend (the label is produced in lines 72–73). Prior to the loop, we print the analytic solution with the shifted time coordinate defined by Eq. (4.10) (see lines 48–53). The results are shown in Fig. 4.2.

The largest time step, $\Delta \tilde{t} = 10^{-3}$ clearly does not work. The initial speed goes through the roof, and the radius becomes by far too large. The solution for $\Delta \tilde{t} = 10^{-4}$ improves somewhat, but the problem is still there. Only the smallest time step, $\Delta \tilde{t} = 10^{-5}$, results in an acceptable approximation with small relative deviation from the analytic solution. However, 200 000 steps are required in this case. It appears to be quite challenging to solve this problem numerically. Can we do better than that?

The Euler method is the simplest method of solving a first order differential equation. There are more sophisticated and more accurate methods. An example is the Runge-Kutta method, which is a higher-order method.[5] This means that the *discretization error* for a finite time step Δt is of the order Δt^n, where $n > 2$. In the

---

[5] A different approach are variable step-size methods. The idea is to adapt the time step based on some error estimate. An example is the Bulirsch-Stoer method. Such a method would be beneficial for treating the initial expansion of a Strömgren sphere. However, with the speed of modern computers,

case of the classic Runge-Kutta method, the error is of order Δt^5. Thus, it is said to be fourth-order accurate and in short called RK4.[6] In contrast, the Euler method is only first-order accurate. This can be seen by comparing the linear approximation (4.6) to the Taylor series expansion

$$r(t + \Delta t) = r(t) + \dot{r}(t)\Delta t + \frac{1}{2}\ddot{r}(t)\Delta t^2 + \dots$$

The Euler method follows by truncating all terms of order Δt^2 and higher. For this reason, the discretization error is also called truncation error. But how can we extend a numerical scheme for an initial value problem to higher order? The first order differential equation (4.7) determines the first derivative $\dot{x}$ for given t and x, but not higher derivatives (do not confuse the order with respect to the truncation error and the order of the differential equation). Nevertheless, higher-order accuracy can be achieved by combining different estimates of the slope $\dot{x}$ at the subinterval endpoints t and $t + \Delta t$ and the midpoint $t + \Delta t/2$.

The RK4 scheme is defined by

$$k_1 = f(t, x)\,\Delta t\,,$$
$$k_2 = f(t + \Delta t/2, x + k_1/2)\,\Delta t\,,$$
$$k_3 = f(t + \Delta t/2, x + k_2/2)\,\Delta t\,,$$
$$k_4 = f(t + \Delta t, x + k_3)\,\Delta t\,,$$

and

$$x(t + \Delta t) = x(t) + \frac{1}{6}\left[k_1 + 2(k_2 + k_3) + k_4\right]\,. \tag{4.12}$$

Here, k_1 is the increment of $x(t)$ corresponding to the Euler method, $k_2/2$ is the interpolated increment for half of the time step, $k_3/\Delta t$ is the corrected midpoint slope based on $x + k_2/2$, and $k_4/\Delta t$ the resulting slope at the subinterval endpoint $t + \Delta t$. Equation (4.12) combines these estimates in a weighted average.

Let us put this into work. Since we need to compute multiple function values for each time step, it is convenient to define the RK4 scheme as a Python function, similar to the numerical integration schemes in Sect. 3.2.2:

```
1  def rk4_step(f, t, x, dt):
2
3      k1 = dt * f(t, x)
4      k2 = dt * f(t + 0.5*dt, x + 0.5*k1)
5      k3 = dt * f(t + 0.5*dt, x + 0.5*k2)
```

---

preference is given to higher-order methods. If you nevertheless want to learn more, you can find a Fortran version of the Bulirsch-Stoer method in [10].

[6]The (local) truncation error is an error per time step. The accumulated error or global truncation error from the start to the endpoint of integration is usually one order lower, i.e. of order Δt^4 in the case of the RK4 method.

```
6       k4 = dt * f(t + dt, x + k3)
7
8       return x + (k1 + 2*(k2 + k3) + k4)/6
```

Here, f() is a Python function corresponding to the right-hand side of Eq. (4.7). In the following, you need to keep in mind that this function expects both the start point t and the value x at the start point as arguments.

We can now compute numerical solutions of the initial value problem for the Strömgren sphere using our implementation of RK4 with different time steps:

```
9   fig = plt.figure(figsize=(6, 4), dpi=100)
10  plt.plot(t, (1 - np.exp(-t+t0))**(1/3),
11          color='red' , label="analytic")
12
13  dt = 1e-3
14  n_steps = 2000
15
16  while dt >= 1e-5:
17      t = np.linspace(0, n_steps*dt, n_steps+1)
18      r = np.zeros(n_steps+1)
19      r[0] = r0
20
21      print("Integrating {:d} steps for dt = {:.0e}".
22            format(n_steps,dt), end=",")
23      for n in range(n_steps):
24          r[n+1] = rk4_step(lambda t, r: (1 - r**3)/(3*r**2),
25                            t[n], r[n], dt)
26
27      # plot the new data
28      plt.plot(t, r, linestyle='--' ,
29              label="Runge-Kutta, $\Delta t$ = {:.1f}".
30              format(dt*ts.to(unit.hr)))
31
32      print(" endpoint deviation = {:.2e}".
33            format(r[-1] - (1 - np.exp(-t[-1]+t0))**(1/3)))
34
35      # decrease time step by a factor of 10
36      dt *= 0.1
37      n_steps *= 10
38
39  plt.legend(loc='lower right')
40  plt.xlabel("$t/t_{\mathrm{s}}$")
41  plt.ylabel("$r/r_{\mathrm{s}}$")
42  plt.ylim(0,1)
43  plt.savefig("stroemgren_rk4.pdf")
```

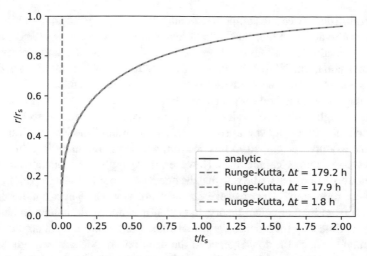

Fig. 4.3 Same plot as in Fig. 4.2 with the Runge-Kutta (RK4) method instead of the Euler method

The Runge-Kutta integrator is called iteratively in lines 24–25. The values passed to `rk4_step()` are `t[n]` and `x[n]`. Moreover, we have to specify the derivate on the right-hand side of the differential equation. For the Strömgren sphere, the mathematical definition reads

$$f(\tilde{t}, \tilde{r}) = \frac{1 - \tilde{r}^3}{3\tilde{r}^2} \; .$$

This can be easily translated into a Python function. Since we need this function only as input of `rk4_step()`, we use an anonymous function rather than a named function (see Sect. 3.2.2). The only pitfall is that our lambda must accept two arguments (listed after the keyword **lambda**), even though only one of them occurs in the expression following the colon (in other words, $f(\tilde{t}, \tilde{r})$ is actually independent of $\tilde{t}$). Of course, the same would apply if we had used a named function. Both arguments are necessary because the function is called with these two arguments in the body of `rk4_step()` (see lines 3–6). You can try to remove the argument `t` in line 24 and take a look at the ensuing error messages.

Compared to the first-order Euler method, the quality of the numerical solutions computed with RK4 improves noticably. As shown in Fig. 4.3, the analytic solution is closely reproduced for d$\tilde{t} = 10^{-4}$ or smaller. The deviation between the numerical and analytic values at the end of integration is printed for each time step in lines 32–33:

```
Integrating 2000 steps for dt = 1e-03, endpoint deviation = 2.08e-01
Integrating 20000 steps for dt = 1e-04, endpoint deviation = 2.23e-04
Integrating 200000 steps for dt = 1e-05, endpoint deviation = 2.26e-07
```

The deviation decreases substantially for smaller time steps. While methods of higher order are more accurate than methods of lower order, they require a larger number of function evaluations. This may be costly depending on the complexity of the differential equation. With four function evaluations, RK4 is often considered a reasonable compromise between computational cost and accuracy. Other variants of the Runge-Kutta method may require a larger or smaller number of evaluations, resulting in higher or lower order. In Appendix B.2, optimization techniques are explained that will enable you to speed up the execution of the Runge-Kutta step.

For the example discussed in this section, we know the analytic solution. This allows us to test numerical methods. However, if we apply these methods to another differential equation, can we be sure that the quality of the solution will be comparable for a given time step? The answer is clearly no. One can get a handle on error estimation from mathematical theory, but often it is not obvious how to choose an appropriate time step for a particular initial value problem. For such applications, it is important to check the convergence of the solution for successively smaller time steps. If changes are small if the time step decreases, such as in the example above, chances are good that the numerical solution is valid (although there is no guarantee in the strict mathematical sense).

4.1.2 Second Order Differential Equations

In mechanics, we typically deal with second order differential equations of the form

$$\ddot{x} = f(t, x, \dot{x}),$$ (4.13)

where $x(t)$ is the unknown position function, $\dot{x}(t)$ the velocity, and $\ddot{x}(t)$ the acceleration of an object of mass m (which is hidden as parameter in the function f). The differential equation allows us to determine $x(t)$ for a given initial position $x_0 = x(t_0)$ and velocity $v_0 = \dot{x}(t_0)$ at any subsequent time $t > t_0$. For this reason, it is called equation of motion.

A very simple example is free fall:

$$\ddot{x} = g,$$ (4.14)

where x is the coordinate of the falling mass in vertical direction. Close to the surface of Earth, $g \approx 9.81 \text{ m s}^{-2}$ and $f(t, x, \dot{x}) = g$ is constant. Of course, you know the solution of the initial value problem for this equation:

$$x(t) = x_0 + v_0(t - t_0) + \frac{1}{2}g(t - t_0)^2.$$ (4.15)

For larger distances from ground, however, the approximation $f(t, x, \dot{x}) = g$ is not applicable and we need to use the $1/r$ gravitational potential of Earth. In this case, the right-hand side of the equation of motion is given by a position-dependent function $f(x)$ and finding an analytic solution is possible, but slightly more difficult. In general, f can also depend on the velocity $\dot{x}$, for example, if air resistance is taken into account. We will study this case in some detail in Sect. 4.2 and you will learn how to apply numerical methods to solve such a problem. An example, where f changes explicitly with time t is a rocket with a time-dependent mass $m(t)$.

To develop the tools we are going to apply in this chapter, we shall begin with another second order differential equation for which an analytic solution is known. An almost ubiquitous system in physics is the harmonic oscillator:

$$m\ddot{x} + kx = 0, \tag{4.16}$$

which is equivalent to

$$\ddot{x} = f(x) \quad \text{where} \quad f(x) = -\frac{k}{m}x. \tag{4.17}$$

The coefficient k is called spring constant for the archetypal oscillation of a mass attached to a spring. You can easily check by substitution that the position function

$$x(t) = x_0 \cos(\omega_0 t), \tag{4.18}$$

where $\omega_0 = \sqrt{k/m}$ is the angular frequency, solves the initial value problem $x(0) = x_0$ and $\dot{x}(0) = 0$. An example from astronomy is circular orbital motion, for which the Cartesian coordinate functions are harmonic oscillations with period $T = 2\pi/\omega$ given by Kepler's third law.

Similar to a body falling under the influence of air resistance, oscillations can be damped by friction. Damping can be modeled by a velocity-dependent term[7]:

$$m\ddot{x} + d\dot{x} + kx = 0, \tag{4.19}$$

where d is the damping coefficient. Hence,[8]

$$f(x, \dot{x}) = -\frac{d}{m}\dot{x} - \frac{k}{m}x, \tag{4.20}$$

This is readily translated into a Python function:

```
1  def xddot(t, x, xdot, m, d, k):
2      """
3      acceleration function of damped harmonic oscillator
4
```

---

[7] An experimental realization would be a ball attached to a spring residing in an oil bath.

[8] As a further generalization, a time-dependent, periodic force can act on the oscillator (forced oscillation). In that case, f depends explicitly on time.

```
5        args: t      - time
6              x      - position
7              xdot - velocity
8              m   - mass
9              d   - damping constant
10             k   - spring constant
11
12       returns: positions (unit amplitude)
13       """
14       return -(d*xdot + k*x)/m
```

In the following, we use a unit system that comes under the name of *arbitrary units*. This is to say that we are not interested in the specific dimensions of the system, but only in relative numbers. If you have the impression that this is just what we did when normalizing radius and time for the Strömgren sphere in the previous section, then you are absolutely right. Working with arbitrary units is just the lazy way of introducing dimensionless quantities. From the programmer's point of view, this means that all variables are just numbers.

The Euler method introduced in Sect. 4.1.1 can be extended to second order initial value problems by approximating the velocity difference over a finite time step Δt as

$$\Delta v = \ddot{x}\Delta t \simeq f(t, x, \dot{x})\Delta t$$

and evaluating $f(t, x, \dot{x})$ at time t to obtain

$$\dot{x}(t + \Delta t) \simeq \dot{x}(t) + \Delta v$$

The *forward* Euler method is then given by the iteration rules

$$x_{n+1} = x_n + \dot{x}_n\Delta t\,, \tag{4.21}$$
$$\dot{x}_{n+1} = \dot{x}_n + f(t_n, x_n, \dot{x}_n)\Delta t \tag{4.22}$$

starting from initial data x_0 and $\dot{x}_0$. This is a specific choice. As we will see, there are others.

To implement this method, we define a Python function similar to `rk4_step()` for first-order differential equations:

```
15   euler_forward(f, t, x, xdot, h, *args):
16       """
17       Euler forward step for function x(t)
18       given by second order differential equation
19
20       args: f - function determining second derivative
21             t - value of independent variable t
22             x - value of x(t)
23             xdot - value of first derivative dx/dt
```

```
24              h - time step
25              args - parameters
26
27      returns: iterated values for t + h
28      """
29      return ( x + h*xdot, xdot + h*f(t, x, xdot, *args) )
```

Compared to `rk4_step()`, there are two differences. First, the updated position x_{n+1} and velocity $\dot{x}_{n+1}$ defined by Eqs. (4.21) and (4.22), respectively, are returned as tuple. Second, `euler_forward()` admits so-called variadic arguments. The following example shows how to use it:

```
30  import numpy as np
31
32  # parameters
33  m = 1.
34  d = 0.05
35  k = 0.5
36  x0 = 10
37
38  n = 1000 # number of time steps
39  dt = 0.05 # time step
40  t = np.arange(0, n*dt, dt)
41
42  # intialization of data arrays for numerical solutions
43  x_fe = np.zeros(n)
44  v_fe = np.zeros(n)
45
46  # initial data for t = 0
47  x_fe[0], v_fe[0] = x0, 0
48
49  # numerical integration using the forward Euler method
50  # parameters are passed as variadic arguments
51  for i in range(n-1):
52      x_fe[i+1], v_fe[i+1] = \
53          euler_forward(xddot, t[i], x_fe[i], v_fe[i], dt,
54                              m, d, k)
```

The difficulty in writing a numerical solver in a generic way is that functions such as `xddot()` can have any number of parameters. In the case of the damped oscillator, these parameters are m, d, and k. For some other system, there might be fewer or more parameters or none at all. We faced the same problem when integrating the Planck spectrum in Sect 3.2.2. In this case, we used a Python lambda to convert a function with an additional parameter (the effective temperature) to a proxy function with a single argument (the wavelength, which is the integration variable). This maintains a clear interface since all arguments are explicitly specified both in the definition and in the calls of a function. A commonly used alternative is argument packing. In our implementation of the forward Euler scheme, `*args` is a placeholder for multiple positional arguments that are not explicitly specified as formal arguments of the function. In other words, we can pass a varying number of arguments whenever the function is called.[9] For this reason, they are also called variadic arguments. You have already come across variadic arguments in Sect. 3.1.1. In the example above, `m`, `d`, and `k` are variadic arguments of `euler_forward()`. Why do we need them here? Because `xddot()` expects these arguments and so we need to to pass them to its counterpart `f()` inside `euler_forward()` (see line 53–54 and line 29 in the definition above).

The `for` loop beginning in line 51 iterates positions and velocities and stores the values after each time step in the arrays `x_fe` and `v_fe`. The arrays are initialized with zeros for a given number of time steps (lines 43–44). The time array `t` defined in line 40 is not needed for numerical integration, but for plotting the solution. The result is shown as dashed line in Fig. 4.4 (the code for producing the plot is listed below, after discussing further integration methods). Compared to the analytic solution for the parameters chosen in our example (solid line), the oscillating shape with gradually decreasing amplitude is reproduced, but the decline is too slow.

Before we tackle this problem, let us first take a look at the computation of the analytic solution. For $x(0) = x_0$ and $\dot{x}(0) = 0$, it can be expressed as

$$x(t) = x_0 e^{-\gamma t} \left[\cos(\omega t) + \frac{\gamma}{\omega} \sin(\omega t) \right] , \qquad (4.23)$$

where $\gamma = d/2m$ and the angular frequency is given by $\omega = \sqrt{\omega_0^2 - \gamma^2}$ (i.e. it is lower than the frequency of the undamped harmonic oscillator). The solution is a damped oscillation with exponentially decreasing amplitude only if $\gamma < \omega_0$. Otherwise, the oscillation is said to be overdamped and $x(t)$ is just an exponential function. We check for this case and ensure that all parameters are positive by means of exception handling in a Python function evaluating $x(t)/x_0$:

```
55  def osc(t, m, d, k):
56      """
57      normalized damped harmonic oscillator
58      with zero velocity at t = 0
59
60      args: t - array of time values
61            m - mass
62            d - damping constant
63            k - spring constant
```

---

[9]Technically speaking, the arguments are packed into a tuple whose name is `args` (you can choose any other name). This tuple is unpacked into the individual arguments via the unpacking operator `*`. To see the difference, you can insert print statements for both `args` and `*args` in the body of `euler_forward()` and execute a few steps.

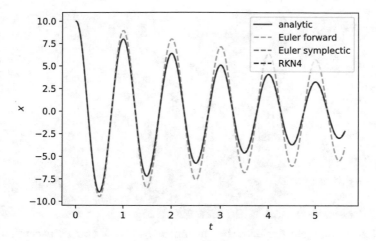

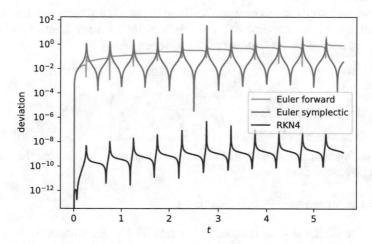

Fig. 4.4 Numerical and analytic solutions of differential equation (4.19) for a damped harmonic oscillator with $m = 1$, $d = 0.05$, $k = 0.5$, $x_0 = 10$, and $\dot{x}_0 = 0$ (top plot). The relative deviations from the analytic solution are shown in the bottom plot

```
64
65      returns: positions (unit amplitude)
66      """
67      try:
68          if m > 0 and d > 0 and k > 0: # positive parameters
69              gamma = 0.5*d/m
70              omega0 = np.sqrt(k/m)
71              if omega0 >= gamma: # underdamped or critical
72                  # frequency of damped oscillation
73                  omega = np.sqrt(omega0**2 - gamma**2)
```

```
74              print("Angular frequency = {:.6e}".
75                      format(omega))
76              return np.exp(-gamma*t) * \
77                      (np.cos(omega*t) +
78                      gamma*np.sin(omega*t)/omega)
79          else:
80              raise ValueError
81      else:
82          raise ValueError
83
84  except ValueError:
85      print("Invalid argument: non-positive parameters
86              or overdamped")
87      return None
```

Invalid arguments are excluded by raising a `ValueError` as exception (see Sect. 3.2.2).

All we need to do for a comparison between Eq. (4.23) and our numerical solution is to call `osc()` for the time sequence defined by the array `t` and multiply the normalized displacements returned by the function with `x0`:

```
88  # analytic solution
89  x = x0*osc(t, m, d, k)
90
91  # relative deviation
92  dev_fe = np.fabs((x - x_fe)/x)
```

When the function `osc()` is called with valid parameters, it prints the frequency of the damped oscillation:

```
Angular frequency = 7.066647e-01
```

The relative deviation from the solution computed with the forward Euler scheme is stored in `dev_fe` and plotted in Fig. 4.4. The accuracy of the numerical solution is clearly not convincing. The average error increases from a few percent to almost 100 % after a few oscillations.

How can we improve the numerical solution? For sure, a smaller time step will reduce the error (it is left as an exercise, to vary the time step and check how this influences the deviation from the analytic solution). But we can also try to come up with a better scheme. It turns out that a small modification of the Euler scheme is already sufficient to significantly improve the solution for a given time step. The forward Euler scheme approximates x_{n+1} and $\dot{x}_{n+1}$ at time $t_n + \Delta t$ by using the slopes $\dot{x}_n$ and $f(t_n, x_n, \dot{x}_n)$ at time t_n. This is an example for an explicit scheme:

values at later time depend only on values at earlier time. What if we used the new velocity $\dot{x}_{n+1}$ rather than $\dot{x}_n$ to calculate the change in position? In this case, the iteration scheme is *semi-implicit*:

$$\dot{x}_{n+1} = \dot{x}_n + f(t_n, x_n, \dot{x}_n)\Delta t \tag{4.24}$$

$$x_{n+1} = x_n + \dot{x}_{n+1}\Delta t \tag{4.25}$$

This scheme is also called *symplectic* Euler method and still a method of first order. Nevertheless, Fig. 4.4 shows a significant improvement over the forward Euler method. Most importantly, the typical error does not increase with time. This is a property of symplectic solvers.[10] Moreover, the relative error is smallest near the minima and maxima of $x(t)$. Owing to divisions by small numbers, the deviation has peaks close to the zeros of $x(t)$. Basically, this causes only small shifts of the times where the numerical solution crosses $x = 0$.

Numerical mathematicians have come up with much more complex methods to achieve higher accuracy. For example, the Runge-Kutta scheme discussed in Sect. 4.1.1 can be extended to second-order differential equations. Moreover, it can be generalized to a class of explicit fourth-order methods, which are known as Runge-Kutta-Nyström (RKN4) methods:

$$\dot{x}_{n+1} = \dot{x}_n + \Delta t \sum_{i=0}^{4} \dot{c}_i f_{ni}, \tag{4.26}$$

$$x_{n+1} = x_n + \Delta t \dot{x}_n + \Delta t^2 \sum_{i=0}^{4} c_i f_{ni}, \tag{4.27}$$

where c_i and $\dot{c}_i$ are method-specific coefficients and f_i are evaluations of the acceleration functions (4.13) at times, positions, and velocities given by

$$f_{n0} = f(t_n, x_n, \dot{x}_n), \tag{4.28}$$

$$f_{ni} = f\left(t_n + \alpha_i \Delta t, \; x_n + \dot{x}_n \alpha_i \Delta t + \Delta t^2 \sum_{j=0}^{i-1} \gamma_{ij} f_{nj}, \; \dot{x}_n + \Delta t \sum_{j=0}^{i-1} \beta_{ij} f_{nj}\right). \tag{4.29}$$

with coefficients α_i, γ_{ij} and β_{ij} for $i = 1, \ldots, 4$ and $j \le i$. These coefficient vector and matrices determine the times, positions, and velocities within a particular time step for which accelerations are evaluated. For an explicit method, the matrices must be lower triangular. You can think of the expressions for positions and velocities in

---

[10]The term symplectic originates from Hamiltonian systems. In fact, the reformulation of the second-order differential equation (4.19) as a set of equations for position and velocity is similar to Hamilton's equations. Symplectic integrators are phase space preserving. This is why the accumulated error does not grow in time.

parentheses and in Eqs. (4.26) and (4.26) as constant-acceleration formulas, assuming different accelerations over the time step Δt. The acceleration values f_{ni} are similar to $k_i / \Delta t$ in the case of the simple Runge-Kutta method introduced in Sect. 4.1.1.

A particular flavor of the RKN4 method (i.e. with specific choices for the coefficients) is implemented in a module accompanying this book. Extract numkit.py from the zip archive for this chapter. In this file you can find the definition of the function rkn4_step() for the RKN4 method along with other methods, for example, euler_step() for the symplectic Euler method. Numerical methods form Sect. 3 are also included. To use any of these functions, you just need to import them.

In the code listed below, you can see how rkn4_step() and euler_step() are applied to the initial value problem for the damped harmonic oscillator. While the code for the Euler method can be easily understood from Eqs. (4.24) and (4.25), we do not go into the gory details of the implementation of RKN4 here. Since the acceleration f_{ni} defined by Eq. (4.29) for a given index i depends on the accelerations f_{nj} with lower indices j (see the sums over j), array operations cannot be used and elements must be computed subsequently in nested **for** loops. This can be seen in the body of rkn4_step() when opening the module file in an editor or using Jupyter. The coefficients are empirical in the sense that they are chosen depending on the performance of the numerical integrator in various tests.

```
93   # apply symplectic Euler and RKN4 schemes
94   from numkit import euler_step, rkn4_step
95
96   x_rkn4 = np.zeros(n)
97   v_rkn4 = np.zeros(n)
98
99   x_se = np.zeros(n)
100  v_se = np.zeros(n)
101
102  x_rkn4[0], v_rkn4[0] = x0, 0
103  x_se[0], v_rkn4[0] = x0, 0
104
105  for i in range(n-1):
106      x_rkn4[i+1], v_rkn4[i+1] = \
107          rkn4_step(xddot, t[i], x_rkn4[i], v_rkn4[i], dt,
108                    m, d, k)
109      x_se[i+1], v_se[i+1] = \
110          euler_step(xddot, t[i], x_se[i], v_se[i], dt,
111                     m, d, k)
112
113  dev_rkn4 = np.fabs((x - x_rkn4)/x)
114  dev_se = np.fabs((x - x_se)/x)
```

Is a numerical scheme as complicated as RKN4 worth the effort? Let us take a look a the solutions plotted with the following code (see Fig. 4.4):

```
15  import matplotlib.pyplot as plt
16  %matplotlib inline
17
18  T = 2*np.pi/7.066647e-01 # period
19
20  fig = plt.figure(figsize=(6, 4), dpi=100)
21  plt.plot(t/T, x, linestyle='-' , color='red' ,
22          label="analytic")
23  plt.plot(t/T, x_fe, linestyle='--' , color='orange' ,
24          label="Euler forward")
25  plt.plot(t/T, x_se, linestyle='--' , color='green' ,
26          label="Euler symplectic")
27  plt.plot(t/T, x_rkn4, linestyle='--' , color='mediumblue' ,
28          label="RKN4")
29  plt.legend(loc='upper right')
30  plt.xlabel("$t$")
31  plt.ylabel("$x$")
32  plt.savefig("oscillator.pdf")
33
34  fig = plt.figure(figsize=(6, 4), dpi=100)
35  plt.semilogy(t/T, dev_fe, linestyle='-', color='orange',
36              label='Euler forward')
37  plt.semilogy(t/T, dev_se, linestyle='-', color='green',
38              label='Euler symplectic')
39  plt.semilogy(t/T, dev_rkn4, linestyle='-' , color='mediumblue',
40              label='RKN4')
41  plt.legend(loc='right')
42  plt.xlabel("$t$")
43  plt.ylabel("deviation")
44  plt.savefig("oscillator_delta.pdf")
```

We have already discussed the two variants of the Euler method (forward and symplectic). If we just look at the plot showing $x(t)$ for the different solvers, it appears that we do not gain much by using RKN4 instead of the symplectic Euler solver. However, the relative deviations reveal that the accuracy of the solution improves by more than six orders of magnitude with the forth-order Runge-Kutta-Nyström method. Although this method is quite a bit more complicated than the Euler method, it requires only five function evaluations vs one evaluation. As explained in Appendix B.1, you can apply the magic command %timeit to investigate how execution time is affected. Apart from that, one can also see that the errors are phase shifted compared to the oscillation (i.e. minimal and maximal errors do not coincide with extrema or turning points of $x(t)$) and the error gradually increase with time. In fact, Runge-Kutta-Nyström solvers are *not* symplectic. As always in numerics, you need to take the requirements of your application into consideration and weigh computational costs against gains.

Exercises

4.1 An important Bernoulli-type equation is the logistic differential equation

$$\dot{x} = kx(1-x),\qquad\qquad (4.30)$$

with the analytic solution

$$x(t) = \frac{1}{1 - e^{-kt}\left(\frac{1}{x(0)} - 1\right)}.\qquad\qquad (4.31)$$

This function has far reaching applications, ranging from population growth over chemical reactions to artificial neuronal networks. Compute numerical solutions of equation (4.30) for $k = 0.5$, $k = 1$, and $k = 2$. Vary the time step in each case and compare to the analytic solution. Suppose the relative error at the time $t = 10/k$ should not exceed 10^{-3}. How many time steps do you need? Plot the resulting graphs for the three values of k.

4.2 Plot the propagation speed $\dot{r}_s$ of the Strömgren sphere discussed in Sect. 4.1.1 relative to the speed of light (i.e. $\dot{r}_s/c$) in the time interval $0 < t/t_s \le 2$. Which consequences do you see for the validity of the model at early time?

4.3 Study the dependence of the error on the time step for the differential equation (4.19), assuming the same initial conditions and parameters as in the example discussed above. Start with an initial time step $\Delta t = T/4$, where $T = 2\pi/\omega$ is the period of the damped oscillation and numerically integrate the initial value problem over two periods. Iteratively decrease the time step by a factor of two until the relative deviation of $x(t = 2T)$ from the analytic solution becomes less than 10^{-3}. List the time steps and resulting deviations in a table for the symplectic Euler and RKN4 methods. Compare the time steps required for a relative error below 10^{-3}. Judging from the order of the truncation error, what is your expectation?

4.2 Radial Fall

Eyewitnesses observed a bright flash followed by a huge explosion in a remote area of the Siberian Taiga on June 30, 1908. The destruction was comparable to the blast of a nuclear weapon (which did not exist at that time). More than a thousand square kilometers of forest were flattened, but fortunately the devastated area was uninhabited. Today it is known as the Tunguska event (named after the nearby river). What could have caused such an enormous release of energy? The most likely explanation is an airburst of a medium sized asteroid several kilometers above the surface of the Earth (in a competing hypothesis, a fragment of a comet is assumed) [11]. In this case, the asteroid disintegrates completely and no impact crater is produced. A sudden explosion can occur when the asteroid enters the lower atmosphere and is rapidly heated by friction.

To estimate the energy released by the impact of an asteroid, let us consider the simple case of radial infall toward the center of the Earth, i.e. in vertical direction (depending on the origin, infalling asteroids typically follow a trajectory that is inclined by an angle relative to the vertical). The equation of motion for free fall in the gravitational potential of Earth is given by

$$\ddot{r} = -\frac{GM_\oplus}{r^2} \tag{4.32}$$

or in terms of the vertical height $h = r - R_\oplus$ above the surface:

$$\ddot{h} = -\frac{GM_\oplus}{(R_\oplus + h)^2} \tag{4.33}$$

Assuming a total energy $E = 0$ (zero velocity at infinity), energy conservation implies

$$v \equiv -\dot{h} = \sqrt{\frac{2GM_\oplus}{R_\oplus + h}}, \tag{4.34}$$

where v is the infall velocity. In particular, the impact velocity at the surface of Earth ($h = 0$) is given by $v_0 = \sqrt{2GM_\oplus/R_\oplus} \approx 11.2$ km/s (use `astropy.constants` to calculate this value). Even for an object following a non-radial trajectory, we can use the kinetic energy $E_{kin} = \frac{1}{2}mv_0^2$ to estimate the typical energy gained by an object of mass m falling into the gravity well of Earth. It is estimated that the asteroid causing the Tunguska event was a solid body of at least 50 m diameter and a mass of the order of 10^5 metric tons. With the radius and density from [11] a kinetic energy of 2.3×10^{16} J is obtained. If such an energy is suddenly released in an explosion, it corresponds to a TNT equivalent of about 5 megatons.[11] The air burst that caused devastation in Siberia in 1908 was probably even stronger because asteroids enter the atmosphere of Earth at higher velocity (the additional energy stems from the orbital motion around the Sun).

To compute the motion of an asteroid in Earth's atmosphere, the following equation of motion has to be solved (we still assume one-dimensional motion in radial direction):

---

[11] One kilogram of the chemical explosive TNT releases an energy of 4184 kJ. The most powerful nuclear weapons ever built produced explosions of several 10 megatons.

$$\ddot{h} = -\frac{GM_\oplus}{(R_\oplus + h)^2} + \frac{1}{2m}\rho_{\text{air}}(h)C_{\text{D}}A\dot{h}^2 \tag{4.35}$$

where $A = \pi R^2$ is the cross section. In addition to the gravity of Earth (first term on the right-hand side), the asteroid experiences air resistance proportional to the square of its velocity (this law applies to fast-moving objects for which the air flowing around the object becomes turbulent). The asteroid is assumed to be spherical with radius R. The drag coefficient for a spherical body is $C_{\text{D}} \approx 0.5$. The dependence of the density of Earth's atmosphere on altitude can be approximated by the barometric height formula:

$$\rho_{\text{air}}(h) = 1.3 \text{ kg/m}^3 \exp(-h/8.4 \text{ km}). \tag{4.36}$$

Equation (4.35) is a non-linear differential equation of second order, which has to be integrated numerically.

Let us first define basic parameters and initial data for the asteroid (based on [11]):

```
1  import numpy as np
2  import astropy.units as unit
3  from astropy.constants import G,M_earth,R_earth
4
5  # asteroid parameters
6  R = 34*unit.m                        # radius
7  V = (4*np.pi/3) * R**3               # volume
8  rho = 2.2e3*unit.kg/unit.m**3        # density
9  m = rho*V                            # mass
```

As start point for the numerical integration of the initial value problem, we choose the height $h_0 = 300$ km. The internal space station (ISS) orbits Earth 400 km above ground. The density of Earth's atmosphere is virtually zero at such altitudes. As a result, we can initially neglect the drag term and calculate the initial velocity v_0 using Eq. (4.34):

```
1  h0 = 300*unit.km
2  v0 = np.sqrt(2*G*M_earth/(R_earth + h0))
```

To apply our Runge-Kutta integrator from numkit, we need to express the second derivative, $\ddot{h}$, as function of t, h, and $\dot{h}$. In the following code, a Python function with the formal arguments required by rkn4_step() and additional parameters for the mass and radius of the asteroid is defined (see Sect. 4.1.2).

```
1  from numkit import rkn4_step
2
3  # drag coefficient
4  c_drag = 0.5
5
6  # barometric height formula
7  def rho_air(h):
8      return 1.3*np.exp(-h/8.4e3)
9
10 # acceleration of the asteroid
11 def hddot(t, h, hdot, m, R):
12
13     # air resistance
14     F_drag =0.5*rho_air(h)*c_drag * np.pi*R**2 * hdot**2
15
16     # gravity at height h
17     g_h = G.value * M_earth.value / (R_earth.value + h)**2
18
19     return -g_h + F_drag/m
```

For the reasons outlined in Sect. 4.1, numerical integration is carried out in a unitless representation (i.e. all variables are simple floats or arrays of floats). To extract numerical values of constants from Astropy, we use the `value` attribute. Parameters of the problem are defined in SI units or converted to SI units.

The integration of the initial value problem for the asteroid is executed by repeatedly calling `rkn4_step()` in a **while** loop until the height becomes non-positive. We use a two-dimensional array called `data` to accumulate time, height, and velocity from all integration steps:

```
20 # initial data
21 data = [[0, h0.to(unit.m).value, -v0.value]]
22
23 # initialization of loop variables
24 t, h, hdot = tuple(data[0])
25 print("Initial acceleration = {:.2f} m/s^2".
26       format(hddot(t, h, hdot, m.value, R.value)))
27
28 # time step
29 dt = 0.1
30
31 while h > 0:
32     h, hdot = rkn4_step(hddot, t, h, hdot, dt,
33                         m.value, R.value)
34     t += dt
35     data = np.append(data, [[t, h, hdot]], axis=0)
```

The array `data` is initialized in line 21 with the initial values, which constitute the first row of a two-dimensional array.[12] Variables for the iteration of t, h, and $\dot{h}$ are defined in line 24, where `tuple()` is used to convert the row `data[0]` into a tuple, which can be assigned to multiple variables in a single statement. The `while` loop starting at line 31 calls `rkn4_step()` for the time step `dt` until h becomes negative. Go through the arguments in lines 32–33 and check their counterparts in the definition of `rkn4_step()`. After incrementing time, the updated variables h and hdot are appended to the data array in line 35. We have already used the function `np.append()` to append elements to a one-dimensional array (see Sect. 3.1.2). In the case of a two-dimensional array, the syntax is more tricky because there are different possibilities of joining multi-dimensional arrays. Here, we want to append a new row. Similar to the initialization in line 21, the row elements have to be collected in a two-dimensional array (as indicated by the double brackets) which is then merged into the existing array. To that end, it is necessary to specify `axis=0`. This axis is in the direction running through rows of an array (corresponding to the first array index) and, thus, indicates that rows are to be appended to rows. If `axis=0` is omitted, `np.append()` will flatten the resulting array, i.e. it is converted into a one-dimensional array with all elements in a sequence (see for yourself what happens to the data array without specifying the axis).[13] Since we do not know the number of time steps in advance, it is convenient to build the data array row by row. However, this means that a new array is created and the complete data from the previous array have to be copied for each iteration (the function's name `append` is somewhat misleading in this regard). For large arrays, this requires too much time and slows down performance of the code. We will return to this issue in Sect. 4.4.

The output produced by the code listed so far is the acceleration $\ddot{h}$ at time $t = 0$:

```
Initial acceleration = -8.94 m/s^2
```

Its absolute value is close to Earth's surface gravity $g = 9.81$ m/s^2. To see how the motion progresses in time, we plot $h(t)$ as a function of t and $\dot{h}$ vs h:

```
36   import matplotlib.pyplot as plt
37
38   plt.figure(figsize=(12,4), dpi=100)
39
40   plt.subplot(121)
41   plt.plot(data[:,0], 1e-3*data[:,1])
42   plt.xlabel("$t$ [s]")
43   plt.ylabel("$h$ [km]" )
44
45   plt.subplot(122)
```

[12] Strictly speaking, the expression with double brackets is a Python list, not a NumPy array. But the list is implicitly converted into an array by `np.append()`.

[13] Arrays can be joined column-wise by using `axis=1` provided that the second array is correctly shaped. A simple example can be found in the notebook for this section.

```
46  plt.plot(1e-3*data[:,1], -1e-3*data[:,2])
47  plt.xlim(h0.value+10,-10)
48  plt.xlabel("$h$ [km]")
49  plt.ylabel("-$\dot{h}$ [km/s]" )
50  plt.savefig("asteroid.pdf")
```

The plot particular variables we need to collect elements column-wise. In Python, there is a simple mechanism for extracting multiple elements from an array, which is called slicing. A slice of an array is equivalent to an index range. For example, data[0,1:2] extracts the two elements with row index 0 and column index running from 1 to 2 (these elements are the height and velocity at time $t = 0$). To obtain the data for the first 10 time steps, you would use the expression data[0:9,:]. If no start and end numbers are specified, the index simply runs through all possible numbers, in this case, from 0 to 2. The first column of the data array, i.e. all time values, is expressed as data[:,0]. This is the first argument of plt.plot() in line 41. Can you identify the slices referring to other variables in the example above? Furthermore, you might find it instructive to experiment with arbitrary slices by printing them.

The plots shown in Fig. 4.5 are produced with the help of plt.subplot(). To arrange two plots in a row, the left subplot is positioned with plt.subplot(121) and the right subplot with plt.subplot(122), where the first digit indicates the number of plots in vertical direction, the second is the number of plots in a row, and the third enumerates the subplots starting from 1. At first glance, the numerically computed solution might appear surprising. The height $h(t)$ is nearly linear although the asteroid is in free fall. The reason is that the free-fall velocities at radial distances $R_\oplus + h_0$ and $R_\oplus$ differ only by a small fraction since the initial height h_0 is small compared to Earth's radius $R_\oplus$ and the asteroid has gained most of its velocity in the long infall phase prior to our arbitrarily chosen initial point $t = 0$. The range along the horizontal axis of the right subplot (velocity vs height) is reverted (see line 47). As a result, the height decreases from left to right corresponding to the progression

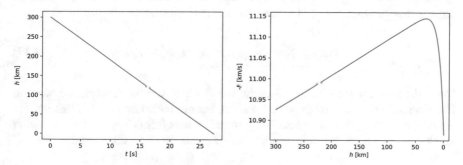

Fig. 4.5 Vertical motion of an asteroid of radius $R = 34$ m through Earth's atmosphere. The left plot shows the altitude h above ground as a function of time. The relation between the asteroid's velocity $\dot{h}$ and h is shown in the right plot

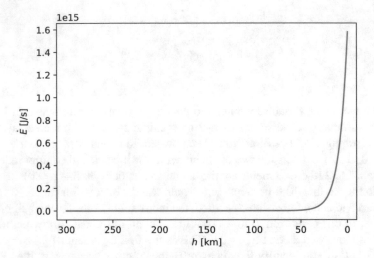

Fig. 4.6 Heating rate caused by air resistance in Earth's atmosphere

of time. As long as the asteroid is nearly in free fall, $|\dot{h}|$ increases gradually, but then the downward speed suddenly drops when the asteroid passes through the lower and much denser layers of the atmosphere. This final phase lasts only a few seconds.

The heating due to air friction equals minus the rate at which kinetic energy is dissipated per unit time. It can be calculated by multiplying the second term in the expression for $\ddot{h}$ with the mass m and velocity $\dot{h}$ (i.e. drag force times velocity):

```
51  def dissipation(h, hdot, m, R):
52      return -0.5*rho_air(h)*c_drag * np.pi*R**2 * hdot**3
```

The dissipation rate is plotted against height in Fig. 4.6 (the code producing this plot can be found in the notebook for this chapter). In the lower layers of Earth's atmosphere, the asteroid decelerates and heating rises dramatically.

How much energy does the asteroid lose in total before it hits the ground? To answer this question, we need to integrate the dissipation rate:

$$E_{\text{diss}}(t) = -\frac{C_{\text{D}} A}{2} \int_0^t \rho_{\text{air}}(h)\dot{h}^3 dt' . \tag{4.37}$$

Since the data for h and $\dot{h}$ are discrete, we can evaluate the function we want to integrate only for $t = n\Delta t$ ($n = 0, 1, 2, \ldots$), but not for arbitrary t. For this reason, we need to modify the numerical integration routines from Chap. 3. The following version of the trapezoidal rule is applicable to an array of function values:

```
53  def integr_trapez(y, h):
54      """
55      numerical integration of a function
56      given by discrete data
57
58      args: y - array of function values
59            h - spacing between x values
60
61      returns: approximate integral
62      """
63      return 0.5*h*(y[0] + 2*np.sum(y[1:-1]) + y[-1])
```

Do not confuse the formal argument h of this function (defined in the local namespace) with the array h, which is the numerical solution of the initial value problem for the asteroid.

As a simple test, let us integrate the function $y = \sin(x)$:

```
64  a, b = 0, np.pi/2
65  n = 10
66  x = np.linspace(a, b, n+1)
67  y = np.sin(x)
```

The array y contains the function values for the chosen subdivision of the integration interval $[0, \pi/2]$. Now we can call `integr_trapez()` with this array and the subinterval width `(b-a)/n` as arguments (compare to Sect. 3.2.2, where `np.sin` is passed as argument):

```
68  integr_trapez(y, (b-a)/n)
```

The output is

```
0.9979429863543572
```

NumPy offers an equivalent library function for numerical integration using the trapezoidal rule:

```
69  np.trapz(y, dx=(b-a)/n)
```

The result agrees within machine precision:

```
0.9979429863543573
```

To compute the fraction of kinetic energy that is dissipated by air resistance, we first extract the data for h and $\dot{h}$ by slicing the full `data` array. These data are used as input to compute discrete values of the dissipation rate, which are in turn integrated with `dt` as subinterval width:

```
70  energy_diss = integr_trapez(
71                      dissipation(data[:,1], data[:,2],
```

```
72                                      m.value, R.value), dt) * unit.J
73  print("Fraction of dissipated energy = {:.2f} %".
74         format(100*energy_diss/energy_kin.to(unit.J)))
```

The ratio of the dissipated energy and the impact energy from above (both energies in units of Joule) is printed as the final result. It turns out that about 5 % of the asteroid's energy are converted into heat (assuming that the asteroid makes it all the way to the surface of Earth):

```
Fraction of dissipated energy = 5.55 %
```

In absolute terms, this is a sizable amount of energy. Since most of this energy is released within a few seconds, it is plausible that the rapid heating may cause the asteroid to disintegrate and burn up, as it likely happened in the Tunguska event. In Exercise 4.5 you can investigate the much stronger impact of air resistance on smaller objects called meteoroids.

Exercises

4.4 Rewrite Simpson's rule (see Sect. 3.2.2) for an array of function values. Compute the dissipated energy for the asteroid discussed in this section and compare to the result obtained with the trapezoidal rule.

4.5 The influence of air resistance increases for smaller objects. Solve the differential equation (4.35) for a meteoroid of radius $R = 25$ cm (rocky objects of size smaller than about one meter are called meteoroids rather than asteroids; when meteoroids enter Earth's atmosphere, they become visible as meteors and their remnants on ground are called meteorites).

1. Plot altitude and velocity of the meteoroid along with the asteroid data from the example above. Interpret the differences.
2. Print the impact time and speed.
3. Estimate roughly when the falling meteoroid becomes a meteor (physically speaking, a meteor is a meteoroid heated to high temperatures by friction).

4.6 The *Falcon 9* rockets produced by SpaceX[14] are famous for their spectacular landing via a so-called *suicide burn*. After deploying the payload at the target orbit, the rocket performs a maneuver that brings it back into Earth's atmosphere where the atmospheric drag controls its maximum velocity. Assuming an equilibrium between gravity and air resistance, the downward acceleration $\ddot{h}$ approaches zero and Eq. (4.35) implies that the terminal velocity for an object of mass m, cross-section A, and drag coefficient c_D is given by

$$v_{max} = \sqrt{\frac{2\,gm}{\rho c_D A}}, \tag{4.38}$$

---

[14]See www.spacex.com/falcon9.

where g is the gravitational acceleration. For instance, in the case of the first stage of a *Falcon 9 Full Thrust Block 5* rocket, the cross-section is nearly circular with a radius of $R = 1.83$ m, a mass of $m_d = 27200$ kg (dry mass without propellant), and $C_D \approx 0.5$.

1. How fast would the rocket hit the sea if it descends to sea level ($h = 0$)?
2. For a safe touchdown, the velocity must be limited to a few meters per second, which is much less than the terminal velocity resulting from atmospheric drag. The rocket fires its thrusters to eliminate any lateral motion before it performs a landing burn to decelerate in vertical direction. Since burning propellant reduces not only the velocity but also the mass of the rocket, which in turn changes the drag term, the computation is more complicated than in the case of an asteroid. For a constant propellant burning rate b, the time-dependent rocket mass can be written as

$$m(t) = m_d + m_p - b\,t \tag{4.39}$$

where m_p is the total propellant mass and m_d the dry mass of the rocket. The modified height equation for a thrust T produced by the rocket engines reads

$$\ddot{h} = -\frac{GM_\oplus}{(R_\oplus + h)^2} + \frac{1}{2\,m(t)}\rho_{air}(h)C_D A\dot{h}^2 - \frac{T}{m(t)}. \tag{4.40}$$

Modify the Python function `hddot()` defined above such that the ignition thrust term $-T/m(t)$ is added for $t \geq t_{ignite}$, where t_{ignite} is the ignition time. Include additional parameters in the argument list.
3. Assume $h_0 = 150$ km and $\dot{h}_0 = 2$ km/s as initial data ($t = 0$). Define an array of ignition times ranging from 50 to 70 s with 0.5 s spacing. For each ignition time, solve Eq. (4.40) for $T = 7.6 \times 10^3$ kN, $m_p = 3.0 \times 10^4$ kg, and $b = 1480$ kg/s. Terminate integration if $h = 0$ is reached or the propellant is exhausted. Determine the height at which the velocity $\dot{h}$ switches signs from negative (descent) to positive (ascend) as a function of t_{ignite} and plot the results.
4. To land the rocket safely, the ignition time must be adjusted such that $\dot{h}$ is nearly zero just at sea level. Estimate the optimal ignition time from the plot (it might be helpful to use `plt.xlim()` to narrow down the range) and interpolate between the nearest date points. Determine the touchdown time, speed, and remaining mass of propellant for the interpolated ignition time.

4.7 How long does it take to fall into a black hole? Well, the answer depends quite literally on the point of view. Suppose a spaceship is plunging from an initial radial distance r_0 toward a black hole of mass M. For simplicity, we assume that the spaceship has zero initial velocity and is accelerated by the gravitational pull of the black hole. We also neglect any orbital angular momentum. Such initial conditions are unrealistic, but our assumptions will suffice for argument's sake. The radial position r at time t is determined by the following differential equation [12]:

$$\frac{dr}{dt} = -c \left(1 - \frac{R_S}{r} \right) \left(\frac{R_S}{r} - \frac{R_S}{r_0} \right)^{1/2} \left(1 - \frac{R_S}{r_0} \right)^{-1/2}, \qquad (4.41)$$

where c is the speed of light and $R_S = 2GM/c^2$ is the Schwarzschild radius of the black hole. In this exercise, we will consider two scenarios: (a) a stellar black hole of mass $M = 10\, M_\odot$ and (b) the supermassive black hole at the center of the Milkyway with $M \approx 4 \cdot 10^6\, M_\odot$. As you can see from the right-hand side of the Eq. (4.41), R_S plays a crucial role. Solve the initial value problem for $r_0 = 100 R_S$ and plot $r(t)$. You will find that the spaceship never crosses the sphere with radius $r = R_S$, which is called the event horizon of the black hole, but appears to hover just above the horizon for eternity.

So is there actually nothing to be feared by someone on board of the spaceship, except being captured in the vicinity of the black hole? In relativity, the progression of time (i.e. the time interval measured by a clock) depends on the observer's frame of reference. The solution following from Eq. (4.41) is what a distant observer far away from the black hole will witness, where the effects of gravity are negligible. For the crew of the spaceship, however, a dramatically different chain of events unfolds. What they read on the starship's clocks is called proper time τ and, for the crew, the rate of change of radial position is given by[15]

$$\frac{dr}{d\tau} = -c \left(\frac{R_S}{r} - \frac{R_S}{r_0} \right)^{1/2}. \qquad (4.42)$$

Solve this equation for the same initial conditions as above. At which time would the spaceship cross the event horizon of the stellar and the supermassive black hole? In the case of the stellar black hole the spaceship will be torn apart by tidal forces even before it reaches the horizon (see Exercise 2.12). Putting aside tidal effects, how long will it take the spaceship until it finally gets crushed by the singularity at $r = 0$? The completely different outcome for the crew of the spaceship is a consequence of gravitational time dilation, which becomes extreme in the case of a black hole.

4.3   Orbital Mechanics

The motion of a planet and a star or two stars around the common center of mass (in astronomy also called barycenter) is governed by Kepler's laws. The properties of the orbits are determined by integrals of motion such as the total energy and orbital angular momentum. Alternatively, we can solve the equations of motion directly. For systems composed of more than two bodies interacting with each other, there is no other way than numerical integration. To begin with, we will apply different

---

[15] In the non-relativistic limit, $\tau \simeq t$ and you can derive Eq. (4.42) from energy conservation $E_{kin} + E_{pot} = 0$.

numerical methods to test whether Keplerian orbits can be reproduced by solving
the initial value problem for two bodies numerically.

As an example, let us compute the orbits of the binary stars Sirius A and B (see
also Sect. 3.1.1). In the center-of-mass reference frame (i.e. the center of mass is at
the origin), we have position vectors

$$\mathbf{r}_1 = -\frac{M_2}{M_1 + M_2}\mathbf{d}, \qquad \mathbf{r}_2 = \frac{M_1}{M_1 + M_2}\mathbf{d}, \tag{4.43}$$

where $\mathbf{d} = \mathbf{r}_2 - \mathbf{r}_1$ is the distance vector between the two stars. The accelerations
are given by

$$\ddot{\mathbf{r}}_1 = \frac{\mathbf{F}_{12}}{M_1}, \qquad \ddot{\mathbf{r}}_2 = \frac{\mathbf{F}_{21}}{M_2}, \tag{4.44}$$

with the gravitational force

$$\mathbf{F}_{12} = -\mathbf{F}_{21} = \frac{GM_1M_2}{d^3}\mathbf{d}. \tag{4.45}$$

To define initial conditions, we make use of the vis-viva equation:

$$v^2 = G\,(M_1 + M_2)\left(\frac{2}{d} - \frac{1}{a}\right), \tag{4.46}$$

where v is the modulus of the relative velocity $\dot{\mathbf{d}}$, G is Newton's constant, and a
is the semi-major axis of the motion of the distance vector $\mathbf{d}(t)$. Remember that
the two-body problem can be reduced to a one-body problem for a body of mass
$\mu = M_1M_2/(M_1 + M_2)$ moving along an elliptic orbit given by $\mathbf{d}(t)$. At the points
of minimal and maximal distance, the velocity vector $\dot{\mathbf{d}}$ is perpendicular to $\mathbf{d}$. At the
periastron, where the two stars are closest to each other, we have

$$\mathbf{r}_1(0) = \left(\frac{M_2}{M_1 + M_2}d_{\mathrm{p}}, 0, 0\right), \tag{4.47}$$

$$\mathbf{r}_2(0) = \left(-\frac{M_1}{M_1 + M_2}d_{\mathrm{p}}, 0, 0\right), \tag{4.48}$$

assuming that the major axes of the ellipses are aligned with the x-axis. For ellipses of
eccentricity e, the periastron distance is given by $d_{\mathrm{p}} = a(1 - e)$ and the corresponding relative velocity, v_{p}, is obtained by substituting d_{p} into Eq. (4.46). By orienting
the z-axis perpendicular to the orbital plane, the orbital velocities at the periastron
can be expressed as

$$\mathbf{v}_1(0) \equiv \dot{\mathbf{r}}_1(0) = \left(0, -\frac{M_2}{M_1 + M_2} v_p, 0\right),   \tag{4.49}$$

$$\mathbf{v}_2(0) \equiv \dot{\mathbf{r}}_2(0) = \left(0, \frac{M_1}{M_1 + M_2} v_p, 0\right).   \tag{4.50}$$

This completes the initial value problem for the two stars.

To proceed with Python, we first define masses and orbitial parameters:

```
1  import numpy as np
2  from scipy.constants import G,year,au
3  from astropy.constants import M_sun
4
5  M1 = 2.06*M_sun.value # mass of Sirius A
6  M2 = 1.02*M_sun.value # mass of Sirius B
7
8  a = 2.64*7.4957*au # semi-major axis
9  e = 0.5914
```

Sirius B is a white dwarf of about one solar mass and Sirius A a more massive main-sequence star. In line 8, the semi-major axis $a \approx 20$ AU is calculated from the distance of the star system from Earth and its angular size. The orbital eccentricity of about 0.6 indicates a pronounced elliptical shape. The orbital period follows from Kepler's third law:

```
1  T = 2*np.pi * (G*(M1 + M2))**(-1/2) * a**(3/2)
2
3  print("Orbital period = {:.1f} yr".format(T/year))
```

Since Sirius A and B orbit each other at relatively large distance, they need years to complete one orbital revolution:

```
Orbital period = 50.2 yr
```

So far, we have solved differential equations for a single function (e.g. the displacement of an oscillator in Sect. 4.1 and the radial coordinate in Sect. 4.2). In the case of the two-body problem, we are dealing with a system of coupled differential equations (4.44) for the vector functions $\mathbf{r}_1(t)$ and $\mathbf{r}_2(t)$. By reformulating the equations of motion as a system of first-order differential equations,

$$\dot{\mathbf{v}}_1 = \frac{GM_2}{|\mathbf{r}_2 - \mathbf{r}_1|^3} (\mathbf{r}_2 - \mathbf{r}_1), \qquad \dot{\mathbf{v}}_2 = \frac{GM_1}{|\mathbf{r}_2 - \mathbf{r}_1|^3} (\mathbf{r}_1 - \mathbf{r}_2),   \tag{4.51}$$

$$\dot{\mathbf{r}}_1 = \mathbf{v}_1, \qquad \dot{\mathbf{r}}_2 = \mathbf{v}_2,   \tag{4.52}$$

it is straightforward to generalize the forward Euler method described in Sect. 4.1.2. First we set the integration interval in units of the orbital period and the number of time steps. Then we initialize arrays for the coordinates and velocity components (the orientation of the coordinate frame is chosen such that the orbits are in the xy-plane and z-coordinates can be ignored):

```
4   n_rev = 3          # number of revolutions
5   n = n_rev*500    # number of time steps
6   dt = n_rev*T/n # time step
7   t = np.arange(0, (n+1)*dt, dt)
8
9   # data arrays for coordinates
10  x1 = np.zeros(n+1)
11  y1 = np.zeros(n+1)
12  x2 = np.zeros(n+1)
13  y2 = np.zeros(n+1)
14
15  # data arrays for velocity components
16  vx1 = np.zeros(n+1)
17  vy1 = np.zeros(n+1)
18  vx2 = np.zeros(n+1)
19  vy2 = np.zeros(n+1)
```

Before proceeding with the numerical integration, we need to assign the initial conditions (4.47) to (4.50) to the first elements of the data arrays:

```
20  # periastron distance and relative velocity
21  d = a*(1 + e)
22  v = np.sqrt(G*(M1 + M2)*(2/d - 1/a)) # vis-viva eq.
23
24  x1[0], y1[0] =  d*M2/(M1 + M2), 0
25  x2[0], y2[0] = -d*M1/(M1 + M2), 0
26
27  vx1[0], vy1[0] = 0, -v*M2/(M1 + M2)
28  vx2[0], vy2[0] = 0,  v*M1/(M1 + M2)
```

Time integration is implemented in the following **for** loop through all time steps:

```
29  alpha = G*M1*M2
30
31  for i in range(n):
32
33      delta_x = x2[i] - x1[i]
34      delta_y = y2[i] - y1[i]
35
36      # third power of distance
37      d3 = (delta_x**2 + delta_y**2)**(3/2)
38
39      # force components
40      Fx = alpha*delta_x/d3
```

```
41        Fy = alpha*delta_y/d3
42
43        # forward Euler velocity updates
44        vx1[i+1] = vx1[i] + Fx*dt/M1
45        vy1[i+1] = vy1[i] + Fy*dt/M1
46        vx2[i+1] = vx2[i] - Fx*dt/M2
47        vy2[i+1] = vy2[i] - Fy*dt/M2
48
49        # forward Euler position updates
50        x1[i+1] = x1[i] + vx1[i]*dt
51        y1[i+1] = y1[i] + vy1[i]*dt
52        x2[i+1] = x2[i] + vx2[i]*dt
53        y2[i+1] = y2[i] + vy2[i]*dt
```

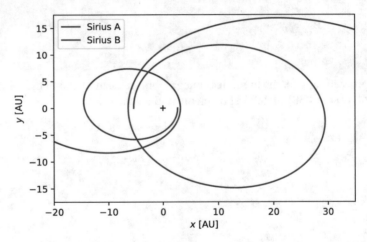

Fig. 4.7 Numerical solution of the two-body problem for the binary stars Sirius A and B computed with the forward Euler method. The center of mass is located at the origin (black cross)

Figure 4.7 shows the resulting orbits (the Python code producing the plot is listed below). Although the shape resembles an ellipse, they solution is clearly not correct. The two stars are gradually drifting outwards and their apastrons (most distant points) are not at the x-axis. As a result, there are no closed orbits. This is in contradiction with the analytic solution of the two-body problem.

```
54  import matplotlib.pyplot as plt
55  %matplotlib inline
56
57  fig = plt.figure(figsize=(6, 6*35/55), dpi=100)
58
59  plt.plot([0], [0], '+k') # center of mass
60  plt.plot(x1/au, y1/au, color='red', label='Sirius A')
```

```
61 | plt.plot(x2/au, y2/au, color='blue', label='Sirius B')
62 |
63 | plt.xlabel("$x$ [AU]")
64 | plt.xlim(-20,35)
65 | plt.ylabel("$y$ [AU]")
66 | plt.ylim(-17.5,17.5)
67 | plt.legend(loc='upper left')
68 | plt.savefig("sirius_forward.pdf")
```

Of course, you know already from Sect. 4.1.2 that the forward Euler method is not suitable to solve dynamical problems. The numerical solution changes remarkably with the following simple modification of lines 50–53:

```
50 |     x1[i+1] = x1[i] + vx1[i+1]*dt
51 |     y1[i+1] = y1[i] + vy1[i+1]*dt
52 |     x2[i+1] = x2[i] + vx2[i+1]*dt
53 |     y2[i+1] = y2[i] + vy2[i+1]*dt
```

After re-running the solver, two closed elliptical orbits are obtained (see Fig. 4.8). Even after three revolutions, there is no noticeable drift. Based on what you learned in Sect. 4.1.2, you should be able to explain how this comes about (a hint is given in the caption of the figure).

How can we further improve the accuracy of the solution? Surely, by using a higher-order scheme. Since the implementation of a Runge-Kutta scheme for systems of differential equations is a laborious task, we make use of library functions for solving initial value problems from SciPy. The wrapper function `scipy.integrate.solve_ivp()` allows you to solve any system of first-order ODEs (higher-order systems can be reformulated as first-order systems) with

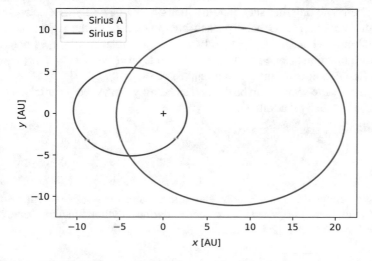

Fig. 4.8 Same as in Fig. 4.7, but computed with a semi-implicit scheme

different numerical integrators.[16] Before applying this function, you need to familiarize yourself with the notion of a state vector. Suppose we have a system of ODEs for N functions $s_n(t)$, where $n \in [1, N]$:

$$\dot{s}_1 = f_1(t, s_1, \ldots, s_N),$$

$$\vdots \qquad\qquad\qquad\qquad (4.53)$$

$$\dot{s}_N = f_N(t, s_1, \ldots, s_N). \qquad\qquad (4.54)$$

This can be written in vector notation as

$$\dot{\mathbf{s}} = \mathbf{f}(t, \mathbf{s}), \qquad\qquad (4.55)$$

where $\mathbf{s} = (s_1, \ldots, s_N)$ and $\mathbf{f} = (f_1, \ldots, f_N)$. Equation (4.55) allows us to compute the rate of change $\dot{\mathbf{s}}$ for any given state $\mathbf{s}$ at time t. In the case of the two-body problem, we can combine Eqs. (4.51), (4.52) into a single equation for an eight-dimensional state vector (again ignoring z-components):

$$\mathbf{s} = \begin{pmatrix} x_1 \\ y_1 \\ x_2 \\ \vdots \\ v_{2x} \\ v_{2y} \end{pmatrix} \quad \text{and} \quad \mathbf{f}(t, \mathbf{s}) = \begin{pmatrix} v_{1x} \\ v_{1y} \\ v_{2x} \\ \vdots \\ -F_{12x}/M_2 \\ -F_{12y}/M_2 \end{pmatrix} \qquad (4.56)$$

It does not matter how the variables are ordered in the state vector, but it is helpful to use mnemonic ordering such that the index n can be easily associated with the corresponding variable. Generally, each component f_n of the right-hand side of Eq. (4.55) may depend on all or any subset of the variables $s_1, \ldots, s_N$. For example, f_1 is only a function of $s_5 = v_{1x}$, while f_8 depends on $s_1 = x_1, s_2 = y_1, s_3 = x_2$, and $s_4 = y_2$ (see definition of the gravitational force (4.45)).

In Python, state vectors can be defined as NumPy arrays. For example, the initial state $\mathbf{s}(0)$ is given by the following array:

```
54  from scipy.integrate import solve_ivp
55
56  init_state = np.array([ x1[0],   y1[0],   x2[0],   y2[0],
57                          vx1[0], vy1[0], vx2[0], vy2[0]])
```

The array elements (initial positions and velocities of the stars) are defined in lines 24–28. To apply `solve_ivp()`, we need to define a Python function that evaluates the right-hand side of Eq. (4.55):

---

[16] See docs.scipy.org/doc/scipy/reference/generated/scipy.integrate.solve_ivp.html.

```
58  def state_derv(t, state):
59      alpha = G*M1*M2
60
61      delta_x = state[2] - state[0]  # x2 - x1
62      delta_y = state[3] - state[1]  # y2 - y1
63
64      # third power of distance
65      d3 = (delta_x**2 + delta_y**2)**(3/2)
66
67      # force components
68      Fx = alpha*delta_x/d3
69      Fy = alpha*delta_y/d3
70
71      return np.array([state[4], state[5], state[6], state[7],
72                       Fx/M1, Fy/M1, -Fx/M2, -Fy/M2])
```

The elements returned by this function are the components of the vector $\mathbf{f}(t, \mathbf{s})$ defined by Eq. (4.56).

The following call of `solve_ivp()` computes the solution for the time interval $[0, 3T]$ starting from `init_state`:

```
73  tmp = solve_ivp(state_derv, (0,3*T), init_state,
74                  dense_output=True)
75  data = tmp.sol(t)
```

The keyword argument `dense_output=True` indicates that the `solve_ivp()` returns a Python function with the name `sol`.[17] This function can be used to compute the values of the solution for any given array of time values via polynomial interpolation. This is what happens in line 75, where the solution is evaluated for the array t defined above (`tmp` is a temporary object containing everything returned by `solve_ivp()`, including `sol()`).

The resulting data can be plotted by means of array slicing, where the row index corresponds to the index of `state` and the column index is running through values at subsequent instants:

```
76  fig = plt.figure(figsize=(6, 6*25/35), dpi=100)
77
78  plt.plot([0], [0], '+k')   # center of mass
79  plt.plot(data[0,:]/au, data[1,:]/au,
80           color='red', label='Sirius A')
81  plt.plot(data[2,:]/au, data[3,:]/au,
82           color='blue', label='Sirius B')
83
84  plt.xlabel("$x$ [AU]")
85  plt.xlim(-12.5,22.5)
86  plt.ylabel("$y$ [AU]")
87  plt.ylim(-12.5,12.5)
```

---

[17] Since a function in Python is an object, a function can be returned by a another function.

```
88  plt.legend(loc='upper left')
89  plt.savefig("sirius_scipy.pdf")
```

The plot is shown in Fig. 4.9. It turns out the orbits are note quite elliptical. The semi-major axes is slowly shrinking. You might find this surprising, as the documentation shows that `solve_ivp()` applies a higher-order Runge-Kutta scheme by default. However, the time step and error tolerances are adjusted such that the integrator computes the solution efficiently with a moderate number of time steps. It is also important to keep in mind that Runge-Kutta methods are not symplectic (see Sect. 4.1.2). Therefore, the error can grow with time. The lesson to be learned here is that numerical library functions have to be used with care. They can be convenient and may offer more sophisticated methods than what you would typically program yourself. But to some degree you need to be aware of their inner workings and tuning parameters to avoid results that do not meet your expectation. In Exercise 4.8, you can further explore `solve_ivp()` and learn how to improve the accuracy of the orbits of Sirius A and B.

To conclude our discussion of orbital mechanics, we will solve a special case of the three-body problem, where two objects are in a close, inner orbit. Together with a third, more distant object, they follows an outer orbit around the common center of mass. Such a configuration is found in the triple star system Beta Persei, which is also known as Algol. Algol Aa1 and Aa2 constitute an eclipsing binary with a period

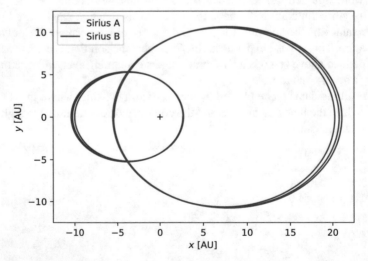

Fig. 4.9 Same as in Fig. 4.8, but computed with the initial value problem solver from SciPy (fifth-order Runge-Kutta scheme with fourth-order error estimator)

of less than three days.[18] This binary and a third star, designated Ab, revolve around each other over a period of 680 days. Here are the orbital parameters of the Algol system:

```python
from scipy.constants import day

M1 = 3.17*M_sun.value # mass of Algol Aa1
M2 = 0.70*M_sun.value # mass of Algol Aa2
M3 = 1.76*M_sun.value # mass of Algol Ab

# inner orbit (Aa1 and Aa2)
T12 = 2.867*day
e12 = 0

# outer orbit (Aa and Ab)
T = 680.2*day
e = 0.227
```

Since the orbital periods are known with higher precision, we compute the semi-major axes of the inner and outer orbits using Kepler's third law, assuming that the binary Aa and the star Ab can be treated as a two-body system with masses $M_1 + M_2$ and M_3. For the inner orbit (Aa1 and Aa2), the third star can be ignored, similar to the Earth-Moon system and the Sun:

```python
from scipy.constants import day

a12 = (T12/(2*np.pi))**(2/3) * (G*(M1 + M2))**(1/3)
a = (T/(2*np.pi))**(2/3) * (G*(M1 + M2 + M3))**(1/3)

print("Inner semi-major axis = {:.2e} AU".format(a12/au))
print("Outer semi-major axis = {:.2f} AU".format(a/au))
```

As expected, the size of the inner orbit is much smaller than the outer orbit (less ten 10 million km vs roughly the distance from the Sun to Mars).

```
Inner semi-major axis = 6.20e-02 AU
Outer semi-major axis = 2.69 AU
```

This allows us to define *approximate* initial conditions. First, we define periastron positions and velocities of Aa1 and Aa2 analogous to Sirius:

```python
M12 = M1 + M2
d12 = a12*(1 - e12)
v12 = np.sqrt(G*M12*(2/d12 - 1/a12))

x1, y1 =  d12*M2/M12, 0
```

---

[18]The term eclipsing binary means that the orbital plane is nearly parallel to the direction of the line of sight from Earth. For such a configuration, one star periodically eclipses the other star, resulting in a characteristic variation of the brightness observed on Earth.

```
26  x2, y2 = -d12*M1/M12, 0
27
28  vx1, vy1 = 0, -v12*M2/M12
29  vx2, vy2 = 0,  v12*M1/M12
```

For the next step, think of Aa1 and Aa2 as a single object of total mass $M_1 + M_2$ (variable `M12`) positioned at the binary's center of mass. By treating the binary and Ab in turn as a two-body system, we have

```
30  d = a*(1 - e)
31  v = np.sqrt(G*(M12 + M3)*(2/d - 1/a))
32
33  x1 += d*M3/(M12 + M3)
34  x2 += d*M3/(M12 + M3)
35
36  x3, y3 = -d*M12/(M12 + M3), 0
37
38  vy1 -= v*M3/(M12 + M3)
39  vy2 -= v*M3/(M12 + M3)
40
41  vx3, vy3 = 0, v*M12/(M12 + M3)
```

In lines 33–34, the x- and y-coordinates are shifted from the center-of-mass frame of Aa1 and Aa2 to the center-of-mass frame for all three stars. The Keplerian velocities at the periastron of the outer orbit are added to the velocities from above (lines 38–39), while Ab moves only with the outer orbital velocity (line 41). As you can see, two further assumptions are applied here. First of all, all three stars are assumed to be simultaneously at their periastrons at time $t = 0$. Generally, periastrons of the inner and outer orbits do not coincide. Second, the plane of the outer orbit of Algol is inclined relative to the plane of the inner orbit. As a result, it would be necessary to treat the motion of the three stars in three-dimensional space. For simplicity's sake, we will ignore this and pretend that the orbits are co-planar.

We solve the equations of motion for the full three-body interactions, i.e. the resulting force acting on each star is given by the sum of the gravitational fields of the other two stars. For example, the equation of motion of Algol Aa1 reads

$$\ddot{\mathbf{r}}_1 = \frac{\mathbf{F}_{12} + \mathbf{F}_{23}}{M_1}, \tag{4.57}$$

where

$$\mathbf{F}_{12} = \frac{GM_1 M_2}{d_{12}^3}\mathbf{d}_{12}, \qquad \mathbf{F}_{13} = \frac{GM_1 M_3}{d_{13}^3}\mathbf{d}_{13}, \tag{4.58}$$

and the displacements vectors are given by

$$\mathbf{d}_{12} = \mathbf{d}_2 - \mathbf{d}_1, \qquad \mathbf{d}_{13} = \mathbf{d}_3 - \mathbf{d}_1. \tag{4.59}$$

Analogous equations apply to Algol Aa2 and Ab. In terms of the state vector of the system, which has twelve components (six positional coordinates and six velocity components), these equations can be implemented as follows.

```
42  def state_derv(t, state):
43      alpha = G*M1*M2
44      beta  = G*M1*M3
45      gamma = G*M2*M3
46
47      delta12_x = state[2] - state[0] # x2 - x1
48      delta12_y = state[3] - state[1] # y2 - y1
49
50      delta13_x = state[4] - state[0] # x3 - x1
51      delta13_y = state[5] - state[1] # y3 - y1
52
53      delta23_x = state[4] - state[2] # x3 - x2
54      delta23_y = state[5] - state[3] # y3 - y2
55
56      # force components
57      F12x = alpha*delta12_x/(delta12_x**2 + delta12_y**2)**(3/2)
58      F12y = alpha*delta12_y/(delta12_x**2 + delta12_y**2)**(3/2)
59
60      F13x =  beta*delta13_x/(delta13_x**2 + delta13_y**2)**(3/2)
61      F13y =  beta*delta13_y/(delta13_x**2 + delta13_y**2)**(3/2)
62
63      F23x = gamma*delta23_x/(delta23_x**2 + delta23_y**2)**(3/2)
64      F23y = gamma*delta23_y/(delta23_x**2 + delta23_y**2)**(3/2)
65
66      return np.array([state[6], state[7],
67                       state[8], state[9],
68                       state[10], state[11],
69                       ( F12x + F13x)/M1, ( F12y + F13y)/M1,
70                       (-F12x + F23x)/M2, (-F12y + F23y)/M2,
71                       (-F13x - F23x)/M3, (-F13y - F23y)/M3])
```

It is left as an exercise to define the initial state vector for the system and to compute and plot the solution using `solve_ivp()`.[19] Figure 4.10 shows the resulting orbits for the time interval $[0, 0.5T]$. Algol Aa1 and Aa2 nearly move like a single object along a Kepler ellipse. In a close-up view the two stars revolve on much smaller orbits around their center of mass (see Fig. 4.11). Combined with the center-of-mass motion, the stars follow helix-like paths. Since the period of the binary's inner orbit is much shorter, a sufficiently small time step has to be chosen (for the orbits shown

---

[19]If this is too time consuming for you, the complete code can be found in the notebook and source files for this chapter.

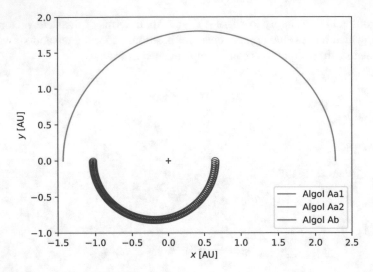

Fig. 4.10 Orbits of three stars similar to the Algol system

in Figs. 4.10 and 4.11 the time step is 0.1 d). This results in a large number of time steps if the equations of motion are integrated over a time interval comparable to the period of the outer orbit. Disparate time scales are a common problem in numerical computation. Often it is not feasible to follow the evolution of a system from the smallest to the largest timescales. In the case of the Algol system, for example, an approximative solution would be to solve the two-body problem for the inner and outer orbits separately. In order to do this, how would you modify the code listed above?

Exercises

4.8 Compare different solvers in `solve_ivp()` for the orbits of Sirius A and B. The solver can be specified with they keyword argument `method`. Moreover investigate the impact of the relative tolerance `rtol`. See the online documentation for details.

4.9 Compute the motion of a hypothetical planet in the Sirius system. Apply the test particle approximation to the planet, i.e. neglect the gravity of the planet in the equations of motion of the stars, while the planets's motion is governed by the gravitational forces exerted by Sirius A and B.

(a) Determine initial data from the apastron of a Kepler ellipse of eccentricity $\epsilon = 0.3$, assuming a single star of mass $M_1 + M_2 \approx 3.08\, M_\odot$ at the center of mass of binary. Under which conditions do you expect this to be a good approximation? Solve the initial value problem for different ratios of the initial distance of the

Fig. 4.11 Close-up view of the inner orbits in the center-of-mass frame of the Algol system

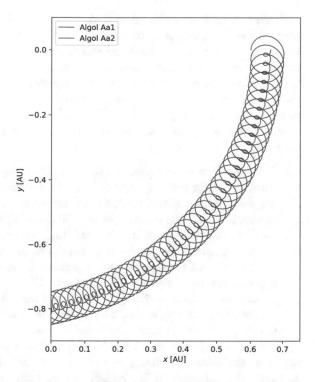

planet from the barycenter and the semi-major axis of the binary star ($a \approx$ 20 AU).[20] In other words, consider cases where the size of the planetary orbit is comparable to the distance between the two stars and where the planet moves far away from the binary. How does binary affect the planet's orbit over several periods and in which respect does it differ from a Keplerian orbit?

(b) Now assume that the planet follows a close circular orbit around Sirius A (i.e. the radius of the orbit is small compared to the separation of the binary). Which approximation can be applied to initialize the orbit in this case? Successively increase the orbital radius from about 0.1 AU to larger values. At which distance from Sirius A destabilizes the planetary orbit within several revolutions (i.e. begins to deviate substantially from the initial elliptical shape)?

4.10 In general, the three body problem cannot be solved analytically. In many cases motions are chaotic, while stable configurations are rare. Consider a system of three stars, each with a mass of 1 $M_\odot$. Intialize the spatial coordinates and velocities of

---

[20] Another parameter is the relative orientation of the major axes of the orbit of the binary and the planet's orbit. You can also vary this parameter if you are interested in its influence.

each star with uniformly distributed random numbers. For the coordinates x, y, and z, generate random numbers in the interval $[-1 \text{ AU}, 1 \text{ AU}]$ using the NumPy function `random.uniform()`, whose first and second argument are the endpoints of the interval from which a random number is to be drawn. By setting the third argument to 3, the function returns three random numbers for the three coordinates. In the same way, generate random velocity components v_x, v_y, and v_z within the interval $[-v_{max}, v_{max}]$, where $v_{max} = \sqrt{GM_\odot/0.01 \text{ AU}}$ is the orbital velocity around a solar mass at a distance of 0.01 AU (this is 10 times the orbital velocity of Earth around the Sun).

(a) Solve the initial value problem for a time interval of at least 10 yr. Plot the pairwise distances d_{12}, d_{13}, and d_{23} between the stars as function of time. Repeat the procedure for several randomly chosen initial conditions and interpret your results. How can you identify bound orbits? (Think about the time behaviour of the distance between two stars in a two-body system.)

(b) The displacement vectors $\mathbf{d}_{12}$, $\mathbf{d}_{13}$, and $\mathbf{d}_{23}$ form a triangle with the three stars at the vertices. Calculate the time-dependent internal angles of the triangle by applying the law of cosines. If the stars are eclipsing (i.e. they are aligned along a line), the triangle will degenerate into a line. As a result, you can detect eclipses by tracking the time evolution of the internal angles. If the configuration approaches an eclipse, the smallest and largest angles will be close to 0 and 180°, respectively. Can you identify such events in your sample?[21]

(c) You can draw a triangle that is similar to the displacement triangle (i.e. a triangle with identical internal angles) by utilizing `Polygon()` from `matplotlib.patches`. Position the first star at the origin, align $\mathbf{d}_{12}$ with the horizontal coordinate axis of the plot and place the third vertex of the triangle such that angles are preserved. What can you deduce from the time evolution of the shape of the triangle? (Optionally, you can return to this exercise after reading the next section. You will then be able to illustrate the time evolution of the three-body system by animating the triangle.)

(d) If you want to optimize the performance of your code to compute larger samples, you can follow the instructions in Appendix B.2.

---

[21] The three-body problem and its rich phenomenology is intriguing for astrophysicists and mathematicians alike. For example, properties of the displacement triangles can be analyzed in an abstract shape space, which can be mapped to a sphere, the so-called shape sphere. The evolution of any three-body system is described by a curve on this sphere. Eclipses live on the sphere's equator. See [13] and references therein.

4.4 Galaxy Collisions

In their eloquent article from the 1980s, Schroeder and Comin [14] came up with a very simple, yet amazingly useful model for interacting disk galaxies. They proposed to treat a disk galaxy   as gravitating point mass surrounded by a disk of test particles representing stars, assuming that all the mass of a galaxy is concentrated in the center and gravitational interactions between stars are negligible. There is no dark matter, no interstellar gas and dust, and stars only experience the gravity of the central mass (or central masses when dealing with a system of two galaxies). Although this is a very crude picture, it is able to reproduce some basic properties seen in interacting galaxies. Of course, it is far from being competitive to a full N-body simulation, which will be the subject of the next section. Since the computation is so much cheaper, it nevertheless serves its purpose as a pedagogical model that can be run in a matter of seconds on any PC or laptop.

In the following, we will reimplement the model of Schroeder and Comin with some extra features in Python.[22] Our version is contained in the module `galcol`, which is part of the zip archive for this chapter. As a first step, we need to initialize a galactic disk. The function `galcol.init_disk()` listed below accepts a dictionary of basic parameters, which can be generated with `galcol.parameters()`:

```
# excerpt from galcol.py
def init_disk(galaxy, time_step=0.1*unit.Myr):
    '''
    initializes galaxy by setting stars in random positions
    and Keplerian velocities half a time step in advance
    (Leapfrog scheme)

    args: dictionary of galaxy parameters,
          numerical time step
    '''

    # width of a ring
    dr = (1 - galaxy['softening'])*galaxy['radius']/ \
        galaxy['N_rings']
    N_stars_per_ring = int(galaxy['N_stars']/galaxy['N_rings'])

    # rotation angle and axis
    norm = np.sqrt(galaxy['normal'][0]**2 +
                   galaxy['normal'][1]**2 +
                   galaxy['normal'][2]**2)
    cos_theta = galaxy['normal'][2]/norm
    sin_theta = np.sqrt(1-cos_theta**2)
    u = np.cross([0,0,1], galaxy['normal']/norm)
    norm = np.sqrt(u[0]**2 + u[1]**2 + u[2]**2)

    if norm > 0:
```

---

[22]The original program GC3D (Gallactic Collisions in 3D) was written in BASIC and later adapted in [4].

```
27        u /= norm # unit vector
28
29        # rotation matrix for coordinate transformation
30        # from galactic plane to observer's frame
31        rotation = \
32            [[u[0]*u[0]*(1-cos_theta) + cos_theta,
33              u[0]*u[1]*(1-cos_theta) - u[2]*sin_theta,
34              u[0]*u[2]*(1-cos_theta) + u[1]*sin_theta],
35             [u[1]*u[0]*(1-cos_theta) + u[2]*sin_theta,
36              u[1]*u[1]*(1-cos_theta) + cos_theta,
37              u[1]*u[2]*(1-cos_theta) - u[0]*sin_theta],
38             [u[2]*u[0]*(1-cos_theta) - u[1]*sin_theta,
39              u[2]*u[1]*(1-cos_theta) + u[0]*sin_theta,
40              u[2]*u[2]*(1-cos_theta) + cos_theta]]
41
42        # print angels defining orientation of galaxy
43        phi = np.arctan2(galaxy['normal'][1],
44                         galaxy['normal'][0])
45        theta = np.arccos(cos_theta)
46        print("Plane normal: ",
47              "phi = {:.1f} deg, theta = {:.1f} deg".\
48              format(np.degrees(phi), np.degrees(theta)))
49
50    else:
51        rotation = np.identity(3)
52
53    galaxy['stars_pos'] = np.array([])
54    galaxy['stars_vel'] = np.array([])
55
56    # begin with innermost radius given by softening factor
57    R = galaxy['softening']*galaxy['radius']
58    for n in range(galaxy['N_rings']):
59
60        # radial and angular coordinates
61        # in center-of-mass frame
62        r_star = R + \
63            dr * np.random.random_sample(size=N_stars_per_ring)
64        phi_star = 2*np.pi * \
65            np.random.random_sample(size=N_stars_per_ring)
66
67        # Cartesian coordinates in observer's frame
68        vec_r = np.dot(rotation,
69                       r_star*[np.cos(phi_star),
70                               np.sin(phi_star),
71                               np.zeros(N_stars_per_ring)])
72        x = galaxy['center_pos'][0] + vec_r[0]
73        y = galaxy['center_pos'][1] + vec_r[1]
74        z = galaxy['center_pos'][2] + vec_r[2]
75
76        # orbital periods and angular displacements
77        # over one timestep
```

```
78      T_star = 2*np.pi * ((G*galaxy['mass'])**(-1/2) * \
79                  r_star**(3/2)).to(unit.s)
80      delta_phi = 2*np.pi * time_step.to(unit.s).value / \
81                    T_star.value
82
83      # velocity components in observer's frame
84      # one half of a step in advance (Leapfrog scheme)
85      vec_v = np.dot(rotation,
86          (r_star.to(unit.km)/time_step.to(unit.s)) * \
87           [(np.cos(phi_star) - np.cos(phi_star-delta_phi)),
88            (np.sin(phi_star) - np.sin(phi_star-delta_phi)),
89            np.zeros(N_stars_per_ring)])
90      v_x = galaxy['center_vel'][0] + vec_v[0]
91      v_y = galaxy['center_vel'][1] + vec_v[1]
92      v_z = galaxy['center_vel'][2] + vec_v[2]
93
94      if galaxy['stars_pos'].size == 0:
95          galaxy['stars_pos'] = np.array([x,y,z])
96          galaxy['stars_vel'] = np.array([v_x,v_y,v_z])
97      else:
98          galaxy['stars_pos'] = \
99              np.append(galaxy['stars_pos'],
00                        np.array([x,y,z]), axis=1)
01          galaxy['stars_vel'] = \
02              np.append(galaxy['stars_vel'],
03                        np.array([v_x,v_y,v_z]), axis=1)
04
05      R += dr
06
07  # units get lost through np.array
08  galaxy['stars_pos'] *= unit.kpc
09  galaxy['stars_vel'] *= unit.km/unit.s
10
11  # typical velocity scale defined by Kepler velocity
12  # at one half of the disk radius
13  galaxy['vel_scale'] = np.sqrt(G*galaxy['mass']/(0.5*R)).\
14                                to(unit.km/unit.s)
```

We follow [14] in subdividing the disk into a given number of rings, `galaxy['N_rings']`. The radial width of each ring `dr` is computed from the outer disk radius `galaxy['radius']` (lines 13–14). The disk also has an inner edge which is specified as fraction of the disk radius. Since this fraction is used to avoid a boundless potential near the center when computing the orbits of test particles, it is called softening factor. While stars are set at constant angular separation in each ring in the original model, we place the stars at random positions. This is done in code lines 62 to 65, where the function `random_sample()` from NumPy's `random` module is used to define the star's radial and angular coordinates within each ring. The sample returned by `random_sample()` is drawn from a uniform distribution in the interval [0, 1] and has to be scaled and shifted to obtain values in the desired range. The number of random values is of course given by the number of stars per ring defined

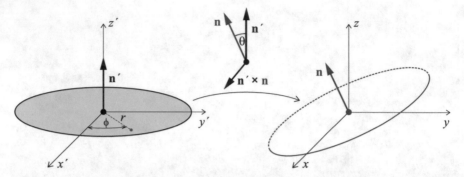

Fig. 4.12 Schematic view of a disk in a coordinate frame aligned with the disk plane and its normal (left) and in another frame with arbitrary orientation relative to the disk (right). The frames are related by a rotation by the angle θ around the axis given by $\mathbf{n}' \times \mathbf{n}$ (middle)

in line 15. The resulting coordinates are defined in a polar coordinate system in the disk plane.

The next problem is transferring the polar coordinates of the stars in the disk to a three-dimensional Cartesian coordinate system (we will refer to this coordinate system as the observer's frame) in which the disk can have arbitrary orientation. The orientation is defined by the normal vector $\mathbf{n} = (n_x, n_y, n_z)$ of the galactic plane, which is defined as a 3-tuple `galaxy['normal']` in the galaxy's dictionary. This requires several steps. First we need to convert the polar coordinates (r, ϕ) into Cartesian coordinates (x', y', z'), where the x' and y' axes are aligned with the disk plane and the z' axis points in the direction perpendicular to the plane (see Fig. 4.12).

$$x' = r \cos \phi, \tag{4.60}$$

$$y' = r \sin \phi, \tag{4.61}$$

$$z' = 0 \tag{4.62}$$

The disk's normal is given by $\mathbf{n}' = (0, 0, 1)$ in this coordinate system. To align the coordinate axes with the observer's frame, the normal direction has to be rotated by an angle θ given by

$$\cos \theta = \mathbf{n}' \cdot \mathbf{n} = n_z , \tag{4.63}$$

assuming that $|\mathbf{n}| = 1$. To ensure that $\mathbf{n}$ is a unit vector, the vector `galaxy['normal']` is divided by its length when defining the variable `cos_theta` in line 21. The rotation axis is then given by the normalized cross product

$$\mathbf{u} = \frac{\mathbf{n}' \times \mathbf{n}}{|\mathbf{n}' \times \mathbf{n}|} , \tag{4.64}$$

as shown in the middle of Fig. 4.12. The cross product can be evaluated by applying the function `np.cross()` (see line 23). Having defined the rotation angle θ and axis $\mathbf{u}$, we obtain the coordinates of a star in the observer's frame, $\mathbf{r} = (x, y, z)^{\mathrm{T}}$,[23] by multiplying $\mathbf{r}' = (x', y', z')^{\mathrm{T}}$ with the rotation matrix R, i.e.

$$\mathbf{r} = \mathsf{R} \cdot \mathbf{r}', \qquad (4.65)$$

where

$$\mathsf{R} = \begin{bmatrix} u_x^2(1 - \cos\theta) + \cos\theta & u_x u_y(1 - \cos\theta) - u_z\sin\theta & u_x u_z(1 - \cos\theta) + u_y\sin\theta \\ u_y u_x(1 - \cos\theta) + u_z\sin\theta & u_y^2(1 - \cos\theta) + \cos\theta & u_y u_z(1 - \cos\theta) - u_x\sin\theta \\ u_z u_x(1 - \cos\theta) - u_y\sin\theta & u_z u_y(1 - \cos\theta) + u_x\sin\theta & u_z^2(1 - \cos\theta) + \cos\theta \end{bmatrix}$$

This expression is is know as Rodrigues' rotation formula. The rotation matrix is defined in lines 31–40 as two-dimensional NumPy array of shape (3,3) provided that the norm of the cross $\mathbf{n}' \times \mathbf{n}$ is positive (line 26) and the right-hand-side of Eq. (4.64) is mathematically defined. If the normal vectors $\mathbf{n}'$ and $\mathbf{n}$ are aligned, the rotation angle is 0 and the rotation matrix is set equal to the identity matrix (line 51).

Equation (4.65) with x', y', and z' substituted by expressions (4.60)–(4.62) is coded in lines 68–74, using np.dot() for the product of a matrix and a vector. Since we work with NumPy arrays of length given by the number of stars per ring, this is done inside the **for** loop through all rings beginning in line 58. The advantage of using dot() is that it can be applied not only to a single vector but also to an array of vectors. (You might find it instructive to figure out the shapes of the variables in lines 68–71 for a particular example). Finally, the x, y, and z coordinates are shifted by the position of the disk center in the observer's frame (lines 72–74).

In the block of code starting at line 85, the orbital velocities of the stars are computed through similar transformations as the positions. Since we are going to solve the equations of motion using a Leapfrog scheme, we need to initialize the velocity of each star by its Keplerian velocity one half of a time step earlier than the initial positions. In the Leapfrog scheme, this is the mean velocity $\mathbf{v}'(t_{-1/2})$ over the time interval $[t_0 - \Delta t, t_0]$:

$$\mathbf{v}'(t_{-1/2}) \simeq \frac{\mathbf{r}'(t_0) - \mathbf{r}'(t_0 - \Delta t)}{\Delta t} \qquad (4.66)$$

For an orbital period T given by Kepler's third law (lines 78–79; see also Sect. 2.2), the angular shift corresponding to Δt is given by (lines 80–81)

$$\Delta\phi = 2\pi \frac{\Delta t}{T}. \qquad (4.67)$$

Thus, the velocity components in the plane of the disk can be expressed as

---

[23]The superscript T indicates that $\mathbf{r}$ is a column vector, which is important in the context of matrix multiplication.

$$v'_x = r\left[\cos\phi - \cos(\phi - \Delta\phi)\right]/\Delta t, \tag{4.68}$$

$$v'_y = r\left[\sin\phi - \sin(\phi - \Delta\phi)\right]/\Delta t, \tag{4.69}$$

$$v'_z = 0 \tag{4.70}$$

Applying the frame rotation, $\mathbf{v} = \mathbf{R} \cdot \mathbf{v'}$, and adding the translation velocity of the disk center yields the velocities in the observer's frame.

In lines 94–103, the computed stellar positions and orbital velocities are accumulated in the arrays `galaxy['stars_pos']` and `galaxy['stars_vel']`, respectively. Once all rings are filled and the loop terminates, the resulting arrays have a shape corresponding to three spatial dimensions times the number of stars in the disk. Finally, Astropy units of kpc and km/s are attached to positions and velocities, respectively. This is necessary because NumPy's `array()` function strips any units from its argument (the reason is explained in Sect. 2.1.3). To compute numerical values that are consistent with the chosen units, radial distance is converted to km and time to s in line 86. It is also noteworthy that we make an exception of our rule of explicitly returning the output of a function here. The position and velocity data are added as new items to the dictionary `galaxy`, which is an argument of `galcol.init_disk()`. Such changes persist outside of a function call.[24] As a result, there is no need to return a complete copy of `galaxy` at the end of the function body.

After having defined initial data, the functions `galcol.evolve_disk()` and `evolve_two_disks()` can be applied to compute the time evolution of a single disk or a pair of disks, respectively. Studying a single disk is left as an exercise (see Exercise 4.11). The definition of `evolve_two_disks()` for the simulation of collisions of galaxies is listed in the following.

```
1   # excerpt from galcol.py
2   def evolve_two_disks(primary, secondary,
3                        time_step=0.1*unit.Myr,
4                        N_steps=1000, N_snapshots=100):
5       '''
6       evolves primary and secondary disk
7       using Leapfrog integration
8
9       args: dictionaries of primary and secondary galaxy,
10              numerical timestep, number of timesteps,
11              number of snapshots
12
13      returns: array of snapshot times,
14               array of snapshots
15               (spatial coordinates of centers and stars)
16       '''
17      dt = time_step.to(unit.s).value
18
19      r_min1 = primary['softening']*primary['radius'].\
```

---

[24]This is an example for Python's call by object reference mentioned in Sect. 3.1.2.

```
20                   to(unit.m).value
21      r_min2 = secondary['softening']*secondary['radius'].\
22                   to(unit.m).value
23
24      N1, N2 = primary['N_stars'], secondary['N_stars']
25
26      # mass, position and velocity of primary galactic center
27      M1 = primary['mass'].to(unit.kg).value
28      X1, Y1, Z1 = primary['center_pos'].to(unit.m).value
29      V1_x, V1_y, V1_z = primary['center_vel'].\
30                          to(unit.m/unit.s).value
31
32      # mass, position and velocity of secondary galactic center
33      M2 = secondary['mass'].to(unit.kg).value
34      X2, Y2, Z2 = secondary['center_pos'].to(unit.m).value
35      V2_x, V2_y, V2_z = secondary['center_vel'].\
36                          to(unit.m/unit.s).value
37
38      # stellar coordinates of primary
39      x = primary['stars_pos'][0].to(unit.m).value
40      y = primary['stars_pos'][1].to(unit.m).value
41      z = primary['stars_pos'][2].to(unit.m).value
42
43      # stellar coordinates of secondary
44      x = np.append(x, secondary['stars_pos'][0].\
45                      to(unit.m).value)
46      y = np.append(y, secondary['stars_pos'][1].\
47                      to(unit.m).value)
48      z = np.append(z, secondary['stars_pos'][2].\
49                      to(unit.m).value)
50
51      # stellar velocities of primary
52      v_x = primary['stars_vel'][0].to(unit.m/unit.s).value
53      v_y = primary['stars_vel'][1].to(unit.m/unit.s).value
54      v_z = primary['stars_vel'][2].to(unit.m/unit.s).value
55
56      # stellar velocities of secondary
57      v_x = np.append(v_x, secondary['stars_vel'][0].\
58                          to(unit.m/unit.s).value)
59      v_y = np.append(v_y, secondary['stars_vel'][1].\
60                          to(unit.m/unit.s).value)
61      v_z = np.append(v_z, secondary['stars_vel'][2].\
62                          to(unit.m/unit.s).value)
63
64      # array to store snapshots of all positions
65      # (centers and stars)
66      snapshots = np.zeros(shape=(N_snapshots+1,3,N1+N2+2))
67      snapshots[0] = [np.append([X1,X2], x),
68                      np.append([Y1,Y2], y),
69                      np.append([Z1,Z2], z)]
70
```

```
71      # number of steps per snapshot
72      div = max(int(N_steps/N_snapshots), 1)
73
74      print("Solving equations of motion for two galaxies",
75            "(Leapfrog integration)")
76
77      for n in range(1,N_steps+1):
78
79          # radial distances from centers with softening
80          r1 = np.maximum(np.sqrt((X1 - x)**2 +
81                                  (Y1 - y)**2 +
82                                  (Z1 - z)**2), r_min1)
83          r2 = np.maximum(np.sqrt((X2 - x)**2 +
84                                  (Y2 - y)**2 +
85                                  (Z2 - z)**2), r_min2)
86
87          # update velocities of stars
88          # (acceleration due to gravity of centers)
89          v_x += G.value * (M1*(X1 - x)/r1**3 +
90                            M2*(X2 - x)/r2**3) * dt
91          v_y += G.value * (M1*(Y1 - y)/r1**3 +
92                            M2*(Y2 - y)/r2**3) * dt
93          v_z += G.value * (M1*(Z1 - z)/r1**3 +
94                            M2*(Z2 - z)/r2**3) * dt
95
96          # update positions of stars
97          x += v_x*dt
98          y += v_y*dt
99          z += v_z*dt
100
101         # distance between centers
102         D_sqr_min = (r_min1+r_min2)**2
103         D_cubed = \
104             (max((X1 - X2)**2 + (Y1 - Y2)**2 + (Z1 - Z2)**2,
105                  D_sqr_min))**(3/2)
106
107         # gravitational acceleration of primary center
108         A1_x = G.value*M2*(X2 - X1)/D_cubed
109         A1_y = G.value*M2*(Y2 - Y1)/D_cubed
110         A1_z = G.value*M2*(Z2 - Z1)/D_cubed
111
112         # update velocities of centers
113         # (constant center-of-mass velocity)
114         V1_x += A1_x*dt; V2_x -= (M1/M2)*A1_x*dt
115         V1_y += A1_y*dt; V2_y -= (M1/M2)*A1_y*dt
116         V1_z += A1_z*dt; V2_z -= (M1/M2)*A1_z*dt
117
118         # update positions of centers
119         X1 += V1_x*dt; X2 += V2_x*dt
120         Y1 += V1_y*dt; Y2 += V2_y*dt
121         Z1 += V1_z*dt; Z2 += V2_z*dt
```

```
122         if n % div == 0:
123             i = int(n/div)
124             snapshots[i] = [np.append([X1,X2], x),
125                             np.append([Y1,Y2], y),
126                             np.append([Z1,Z2], z)]
127
128
129             # fraction of computation done
130             print("\r{:3d} %".format(int(100*n/N_steps)), end="")
131
132     time = np.linspace(0*time_step, N_steps*time_step,
133                        N_snapshots+1, endpoint=True)
134     print(" (stopped at t = {:.1f})".format(time[-1]))
135
136     snapshots *= unit.m
137
138     return time, snapshots.to(unit.kpc)
```

Parameters and initial data of the two galaxies are defined by the arguments primary and secondary, which have to be dictionaries prepared by galcol.init_disk(). For the numerical solver in the **for** loop through all timesteps (the total number of steps is specified by the optional argument N_steps), parameters and data from the dictionaries are converted to simple float values in SI units (lines 17–62). For example, the coordinates X1, Y1, and Z1 of the center of the primary galaxy in units of meters are initialized in line 28. The conversion from Astropy objects to numbers avoids some overhead in the implementation and improves efficiency of the computation, while we have the full flexibility of using arbitrary units in the input and output of the function.

The positions of stars in both disks are joined into arrays x, y, and z via np.append() in lines 39–49. As a result, the length of the three coordinate arrays equals the total number of stars in the primary and secondary disks. This allows us to apply operations at once to all stars. The coordinates are updated for each time step in lines 97–99, where the velocity components are computed from the accelerations in the gravitational field of the two central masses M_1 and M_2 (lines 89–94):

$$\mathbf{x}(t_{n+1}) = \mathbf{x}(t_n) + \mathbf{v}(t_{n+1/2})\Delta t, \tag{4.71}$$

$$\mathbf{v}(t_{n+1/2}) = \mathbf{v}(t_{n-1/2}) + \mathbf{a}(t_n)\Delta t, \tag{4.72}$$

where

$$\mathbf{a}(t_n) = \frac{GM_1}{r_1^3(t_n)}\left[\mathbf{X}_1(t_n) - \mathbf{x}(t_n)\right] + \frac{GM_2}{r_2^3(t_n)}\left[\mathbf{X}_2(t_n) - \mathbf{x}(t_n)\right] \tag{4.73}$$

is the gravitational acceleration of a test particle (star) at position $\mathbf{x}(t_n)$ at time $t_n = t_0 + n\Delta t$. The distances $r_{1,2}(t_n)$ from the two central masses are given by (for brevity, time dependence is not explicitly written here):

$$r_{1,2} = \sqrt{(X_{1,2} - x)^2 + (Y_{1,2} - y)^2 + (Y_{1,2} - y)^2}. \tag{4.74}$$

However, in the corresponding code lines 80–85 distances are given by the above expression only if the resulting distances are greater than some minimal distances r_min1 and r_min2. NumPy's maximum() function compares values element-wise. In this case, it compares each element of the array r1 to r_min1 and returns the larger value (and similarly for r2). The variables r_min1 and r_min2 are defined in terms of the disk's softening factors in lines 19 to 22. This means that in the close vicinity of the gravitational centers, the potential is limited to the potential at the minimal distance. Otherwise stellar velocities might become arbitrarily high, resulting in large errors.

Of course, not only the stars move under the action of gravity, but also the two central masses M_1 and M_2. Since the stars are treated as test masses (i.e. the gravity of the stars is neglected), it is actually a two-body problem that needs to be solved for the central masses. We leave it as an exercise to study the implementation in lines 101–121 (see appendix of [4] for a detailed description). The expressions for the velocity updates of the secondary in lines 114–116 follow from momentum conservation.

The data for the positions of the centers and stars are stored in snapshots, a three-dimensional array that is initialized with np.zeros() in line 66. Since we want to record the evolution of the system, we need to store a sufficient number of snapshots to produce, for instance, an animation of the two disks. The most obvious choice would be to store the data for all timesteps. However, this would result in a very large array consuming a lot of memory. Moreover, the production of animations becomes very time consuming if the total number of frames is too large. The number of snapshots can be controlled with the optional argument N_snapshots. The shape of the array snapshots specified in the call of np.zeros() is the number of snapshots plus one (for the initial data) times the number of spatial dimensions (three) times the total number of stars plus two (for the two centers). The initial positions of centers and stars defines the first snapshot snapshots[0], which is a two-dimensional subarray (lines 67–69). In principle, we could gradually extend snapshots by applying np.append() for each subsequent timestep, but this involves copying large amounts of data in memory. As a consequence, the program would slow down considerably with growing size of snapshots. For this reason, it is advantageous to define large arrays with their final size (you can check the size in the examples discussed below) and to successively set all elements in the aftermath of filling them with zeros. For the default values defined in line 4, one snapshot will be produced after every cycle of 100 timesteps (the number of steps per snapshot is assigned to the variable div in line 72). Consequently, we need to check if the loop counter n (the current number of timesteps in the loop body) is a multiple of 10. This is equivalent to a zero remainder of division of n by 10. In Python, the remainder is obtained with % operator. Whenever this condition is satisfied (line 123), the position data are assigned to the snapshot with index given by n/div (lines 124–127). Remember that an array index must be an integer. Since Python performs divisions in floating point arithmetic, we need to apply the conversion function int() when calculating the snapshot index. After the loop over all timesteps has finished, snapshots is multiplied by unit.m and converted to kpc before it is returned. (Do you see why can we not just multiply by unit.kpc?)

The function also returns an array `time` with the instants of time corresponding to the snapshots, which can be used to label visualizations of the snapshots.

Let us do an example:

```
139  import galcol
140  import astropy.units as unit
141
142  galaxies = {
143      'intruder' : galcol.parameters(
144          # mass in solar masses
145          1e10,
146          # disk radius in kpc
147          5,
148          # coordinates (x,y,z) of initial position in kpc
149          (25,-25,-5),
150          # x-, y-, z-components of initial velocity in km/s
151          (-75,75,0),
152          # normal to galactic plane (disk is in xy-plane)
153          (0,0,1),
154          # number of rings (each ring will be randomly
155          # populated with 1000/5 = 200 stars)
156          5,
157          # total number of stars
158          1000,
159          # softening factor defines inner edge of disk
160          0.025),
161      'target' : galcol.parameters(
162          5e10, 10, (-5,5,1), (15,-15,0), (1,-1,2**0.5),
163          10, 4000, 0.025),
164  }
```

First, we need to import the modules `galcol` and `astropy.units`. Then an intruder and a target galaxy are defined as items in the dictionary named `galaxies`. To help you keep an overview of the parameters, comments are inserted in the argument list of `galcol.parameters()` for the intruder (to see the definition of the function `parameters()`, open the file `galcol.py` with an editor). Compared to the target, the intruder has a five times smaller mass and is also smaller in size. It approaches the intruder with a relative velocity of 128 km/s from an initial distance of 30 kpc under an angle of 45° in the xy-plane and a separation of 6 kpc in transversal direction. The initial positions and velocity vectors of the two disks are chosen such that their center of mass resides at the origin of the coordinate system (since we are dealing with a two-body problem with central forces, the center of mass is stationary).

Test particles are produced by invoking `galcol.disk_init()` for both disks:

```
165  galcol.init_disk(galaxies['intruder'])
166  galcol.init_disk(galaxies['target'])
```

If you output the contents of `galaxies['intruder']` after the call above, you will find arrays containing the particle's initial positions and velocities under the keywords `'stars_pos'` and `'stars_vel'`, respectively (with different numbers produced by your random number generator):

```
{'mass': <Quantity 1.e+10 solMass>,
 'radius': <Quantity 5. kpc>,
 'center_pos': <Quantity [ 25., -25.,  -5.] kpc>,
 'center_vel': <Quantity [-75.,  75.,   0.] km / s>,
 'normal': (0, 0, 1),
 'N_rings': 5,
 'N_stars': 1000,
 'softening': 0.025,
 'stars_pos': <Quantity [[ 25.27909275,  24.83823236,  25.05526353, ...,  26.70027151,
          28.50078551,  22.24242959],
         [-25.164745  , -24.60920606, -24.43662745, ..., -21.34249645,
          -22.12864831, -20.94120214],
         [ -5.        ,  -5.        ,  -5.        , ...,  -5.
          -5.        ,  -5.        ]] kpc>,
 'stars_vel': <Quantity [[ 127.78537167, -374.04767989, -348.54051012, ...,
          -168.58250706, -136.72561385, -152.49071953],
         [ 377.39033859,  -35.49029092,  108.72755079, ...,
          118.65298369,  150.42575762,   22.46279061],
         [   0.        ,    0.        ,    0.        , ...,
            0.        ,    0.        ,    0.        ]] km / s>,
 'vel_scale': <Quantity 131.16275798 km / s>}
```

The next step is to compute the time evolution of the combined system of intruder and target, starting for the initial data produced above:

```
167  t, data = galcol.evolve_two_disks(
168                  galaxies['target'], galaxies['intruder'],
169                  N_steps=10000, N_snapshots=500,
170                  time_step=0.05*unit.Myr)
```

While the integration proceeds, progress is printed in percent of the total number of time steps, which is 10000 in this case. This is achieved by a print statement in the main loop (see line 130 in the code listing of `evolve_two_disks()`); the format option `end=" "` prevents a new line after each call and, owing to the carriage return `"r"`, printing starts over at the beginning of the same line). The function completes with

```
Solving equations of motion for two galaxies (Leapfrog integration)
100 % (stopped at t = 500.0 Myr)
```

The final time $t = 500$ Myr follows from the chosen timestep, $\Delta t = 0.05$ Myr, times the total number of timesteps.

Various options for visualization are available in `galcol` (in the exercises, you have the opportunity to explore their capabilities). For example, to produce a three dimensional scatter plot of the stars for a particular snapshot, you can use

```
171  i = 100
172  galcol.show_two_disks_3d(data[i,:,:],
173                           galaxies['target']['N_stars'],
174                           [-15,15], [-15,15], [-15,15], t[i],
175                           'two_disks')
```

This call was used to produce the upper left plot for $t = 100$ Myr in Fig. 4.13. From the parameters in lines 169–170 you an calculate that the snapshot index i simply corresponds to the time in Myr. In the call above, the slice data[i,:,:] for the snapshot with index i is passed as first argument of show_two_disks_3d(). The second argument allows the function to infer the number of stars in the two disks, which is needed to display stars belonging to the intruder and target galaxies in blue and red, respectively. The following arguments set the x, y, and z-range of the plot in units of kpc. The snapshot time $t[i]$ is required for the label on top of the plot (if it is omitted, no label is produced). The last argument is also optional and specifies the prefix of the filename under which the plot is saved (the full filename is composed from the prefix and the snapshot time). The rendering is based on the scatter() function from pyplot. In contrast to a surface plot (see Sect. 3.2.1), a scatter plot shows arbitrarily distributed data points. The function scatter() can also be used for two-dimensional plots, i.e. points in a plane. As you can see from the source code in the file galcol.py, we use Axes3D from mpl_toolkits.mplot3d to create three-dimensional coordinate axes.

Figure 4.13 shows a sequence of plots illustrating the evolution of the two disks ranging from $t = 100$ Myr, where the intruder is still approaching the largely unperturbed target, to 450 Myr, where the remnants of the galaxies are moving apart (the centers of mass follow hyperbolic trajectories). The intruder is strongly disrupted during its slingshot motion through the potential well of the target's larger central mass. The plots in the middle show that a large fraction of the intruder's stars are ejected, while the intruder triggers spiral waves in the target disk through tidal forces. Although some stars in the outer part of the target disk are driven away from the center, the effect is less dramatic because the stars are bound more tightly. There is also function anim_two_disks_3d() for producing an animation of the snapshots in the module file galcol.py . While we do not discuss the details here, you can find an example in the online material for this chapter. The animation is saved in MP4 format and you can view it with common movie players.

The Whirlpool Galaxy M51 is a well known example for a close encounter of two galaxies (see also Sect. 5.4). There are other types, for example, the Cartwheel galaxy (see Exercise 4.13). In Fig. 4.14, you can see an optical image of a pair of galaxies with prominent tidal tails made by the Hubble Space Telescope (HST). Our simulation resembles such galaxies although the underlying model is very simple and completely ignores the gas contents of galaxies and the dark matter halos (see also the discussion in [14]). However, one should keep in mind that it is unrealistic in some important aspects:

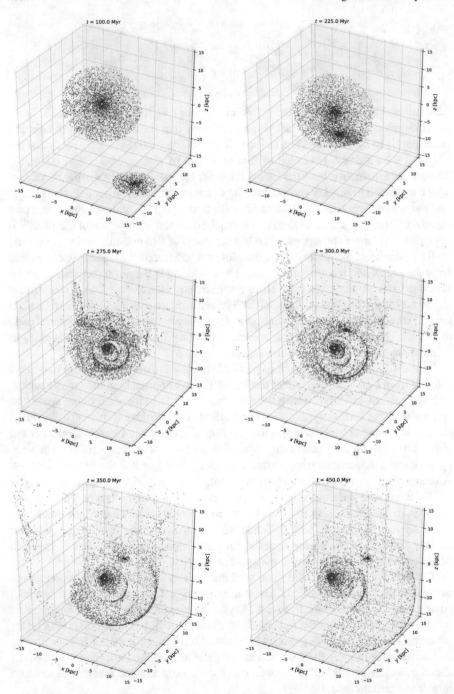

Fig. 4.13 Different stages of the collision of two galaxies (time in Myr is indicated on top of each plot). Stars of the target galaxy are shown in blue, stars of the intruder in red. The target disk has an inclination of 45° relative to the horizontal (*xy*) plane

Fig. 4.14 Hubble image of interacting galaxies Arp 87. Image credit: NASA, ESA, and The Hubble Heritage Team (STScI/AURA)

- The collision of real galaxies is not governed by two-body dynamics. The dark matter halos with their embedded disks of baryonic matter can even merge into a single galaxy.[25] It is believed that galaxies typically undergo several merges in the course of cosmic history.
- The ejection of stars through the slingshot effect, which is caused by strong acceleration during the close flyby of a gravitating center, is prone to numerical errors. To compute the trajectories more accurately a very low timestep would be required, which in turn would substantially increase the computing time. You can investigate the impact of the timestep on different trajectories in Exercise 4.14.

Exercises

4.11 Compute the evolution of an isolated disk in the xy-plane centered at $(0, 0, 0)$ using `galcol.evolve_disk()`.

1. First consider the case where the center is at rest. Visualize the time evolution of the disk with `galcol.anim_disk_2d()`. Does the behaviour of the disk meet your expectation?
2. Since there are no perturbations of orbital motion by a second disk, the stars should follow circular Keplerian orbits. The function `galcol.show_orbits()` allows you to plot the numerically computed orbits of individual stars (the indices of these stars are passed as elements of an array to the function). Choose stars in the different rings and compare their motion over a given interval of time. Compute the disk evolution with different numerical timesteps. How are the orbits affected by the timestep, particularly in the innermost ring?

---

[25]The term baryonic refers to elementary particles in the atoms of which gas and stars are composed.

4.12 The effect of the collision of two galaxies depends mainly on their relative velocity and the impact parameter b, which is defined as the perpendicular distance between the path of the intruder galaxy from infinity and the center of the target galaxy. If the separation of the two galaxies at time $t = 0$ is large enough, it can be assumed that the intruder is nearly unaffected by the gravity of the target and moves along a straight line through its center in the direction of its initial velocity vector. The impact parameter is then given by the normal distance of this line to the center of the target.

1. Calculate b in kpc for the scenario discussed in this section. Vary the impact parameter by changing the initial position of the intruder. Compute the resulting evolution of the two disks and interpret the results.
2. What is the effect of the relative velocity of the intruder and the target for a given impact parameter?
3. The relative orientation of the disks and the mass ratio also play a role in the interaction process. Investigate for one of the scenarios from above orientations of the target disk ranging from $\theta = 0°$ (planes of target and intruder are parallel) to $90°$ (target perpendicular to intruder) and compare the mass ratios $1 : 5$ (example above) and $1 : 1$ (equal masses). Discuss how the central masses and stars are affected by these parameters.

4.13 The head-on collision of two disks can result in a Cartwheel-like galaxy [4, 14]. The name refers to the large outer ring which gives the galaxy the appearance of a wagon wheel. In this case, the intruder moves in z-direction toward the target and its normal is aligned with the direction of motion. The plane of the target disk is parallel to the intruder's disk. Vary the relative velocity and the impact parameter. Can you produce a post-collision galaxy of a similar shape as the Carthwheel Galaxy?

4.14 Analyze trajectories of ejected stars in some of the simulations from Exercise 4.12 or 4.13. Since you cannot predict which star in the initial disks will be ejected, take random samples of stars and plot their orbits with `galcol.show_orbits_3d()`.

1. Compute and plot the time-dependent specific orbital energy

$$\epsilon(t) = \frac{1}{2}v(t)^2 - \frac{GM_1}{r_1(t)} - \frac{GM_2}{r_2(t)}, \tag{4.75}$$

   for your sample of stars. The distances $r_{1,2}$ from the two galaxy centers are defined by Eq. (4.74). To compute the kinetic energy per unit mass, $v^2/2$, you need to modify `galcol.evolve_two_disks()` such that the position and velocity data are returned for each snapshot. Compare ejected stars to stars that remain bound to the target galaxy and describe the differences. Why is $\epsilon(t)$ in general not conserved?
2. How sensitive is $\epsilon(t)$ to the numerical time step?

4.15 Implement an object-oriented version of `galcol.py` by defining a disk-galaxy class that has basic parameters and state data for all particles as attributes. See Appendix A for an introduction to classes. If you are interested in learning more about object-oriented programming in Python, we encourage you to study advanced textbooks on the subject. Here comes a teaser for mastering object-oriented programming: Since the galaxy collision model is derived from the two-body problem, you can alternatively make use of inheritance and define your galaxy class as a subclass of the class `Body` introduced in Appendix A.

4.5 Stellar Clusters

With the advent of increasingly powerful supercomputers, many-body gravitational systems could be studied in unprecedented detail where earlier generations had to rely on statistical analysis. Today, even typical desktop computers are capable of simulating thousands of bodies while the largest computing facilities can deal with millions or - utilizing numerical approximations - billions of bodies.

The most important application in astrophysics is the dynamics of dark matter halos in which galaxies, groups of galaxies, or clusters consisting of thousands of galaxies are embedded. The dark matter is treated as collisionless gas, where particles interact only via gravity. In this case, the term particle does not mean an atom or elementary particle, it just refers to an arbitrary point mass. For a total number of N particles, $\sim N^2$ interactions have to be computed to determine the instantaneous accelerations of all particles directly. Since the particle positions change in time, the computation of interactions has to be carried out for a larger number of time steps. This is an intractable task for typical particle numbers $N \sim 10^9$ or even larger in modern N-body simulations. For this reason, approximations are used. A simple example is the algorithm for galaxy collisions in the previous section, where the gravity of stars ("test particles") was neglected and the acceleration of each particle has only two terms (see Eq. 4.73). Without neglecting the stellar masses, we would need to sum over N terms, where N is the total number of stars plus the two central masses.

For a moderate number of particles ($N \lesssim 10^3$), the solution of the N-body problem via direct summation over all interactions is feasible on a typical personal computer. However, it advisable to use compilable code written in languages such as C or Fortran for this kind of computation. Since the source code is turned into a machine-readable program that can be executed without on-the-fly translation by an interpreter, it usually performs numerical computations more efficiently. Alternatively, compilable Python modules can be used (see Appendix B.2). The difference is insignificant for most of the applications in this book, but N-body dynamics is definitely an exception.

In the following, we analyse the motion of stars in a globular cluster. Globular clusters are stellar clusters of spherical shape. They are found in the halos of a galaxies and contain hundreds of thousands of stars in a fairly small volume ([3], Sect. 17.3). Consequently, the density of stars is much higher than in galactic disks.

Fig. 4.15 Image of M 13
taken by Walter Baade with
the 1 m reflector of Hamburg
Observatory in 1928. The
image shows the central
section of a photographic
plate, which can be
downloaded in FITS format
(see Chap. 5) from the
Digital Plate Archive of
Hamburg Observatory
(plate-archive.hs.uni-hamburg.de).
Such plates, which are made
of glass covered by a
light-sensitive emulsion,
were widely used in
telescopes before digital
CCD cameras became
available in the 1980s
(Credit: Plate Archive of
Hamburg Observatory,
University of Hamburg)

A prominent globular cluster belonging to the Milkyway Galaxy is Messier 13 (see Fig. 4.15). As an example, we use data from an N-body simulation of a downsized cluster with 1000 stars.[26] The evolution of the cluster was computed over a period of 100 million years and snapshots of the stellar positions and velocities were recorded every 200,000 years, resulting in 500 output files. The files can be downloaded as part of data_files.zip from uhh.de/phy-hs-pybook. After you have downloaded the archive, extract it into the work directory containing your Python source code or Jupyter notebooks.

Each output file contains formatted positions and velocities of all stars in a table that can be loaded from disk and stored in a two-dimensional NumPy array by using np.loadtxt(). Since we have a large number of output files, we collect the complete data in a three-dimensional data array, where the first index represents time, the second index indentifies the star, and the third index runs through the different variables for each star:

---

[26]The simulation code is published by Marcel Völschow on github.com/altair080/nbody.

```
1   import numpy as np
2
3   n_files = 501
4
5   for n in range(n_files):
6       # load n-th snapshot from file
7       snapshot = np.loadtxt(
8           "data_files/nbody/output_{:d}.dat".format(n),
9           dtype='float64')
10
11      if n == 0:
12          # create data array with first snapshot as element
13          data = np.array([snapshot])
14      else:
15          # append further snapshots to data array
16          data = np.append(data, [snapshot], axis=0)
```

The data files are loaded in lines 7–9, where it is assumed that the files can be found in the subfolder data_files/nbody/. If the data files are located in some other folder, you need to adjust the path accordingly. The syntax used above works on Linux and macOS systems. If you work on a Windows computer, you need to replace / by the backslash character \. The files are numbered from 0 to 500. For each iteration in the loop, the file name is generated from the loop index n via .format(n). The keyword argument dtype='float64' indicates that the numbers in the file are to be interpreted as floating point numbers of double precision (this is the standard type for floats in Python). The contents loaded from a particular file is first stored in snapshot (a two-dimensional array) and then appended to the three-dimensional array data. Let us check whether data has the expected shape:

```
17  data.shape
```

outputs

```
    (501, 1000, 8)
```

We have 501 snapshots (including the initial data) for 1000 stars and 8 entries per snapshot and star.

To take a look at the initial the distribution of stellar masses, we can load the required data via array slicing and produce a histogram:

```
18  import matplotlib.pyplot as plt
19  %matplotlib inline
20
21  plt.figure(figsize=(6,4), dpi=100)
22  plt.hist(data[0,:,1], 20, histtype='step', lw=2)
23  plt.xlabel('$M/M_\odot$')
24  plt.ylabel('$N$')
25  plt.savefig('cluster_hist_mass.pdf')
```

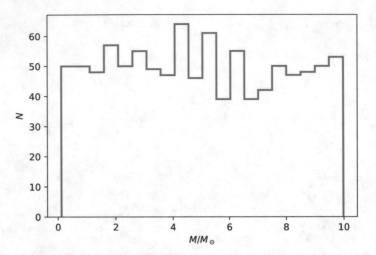

Fig. 4.16 Histogram of stellar masses in the cluster

The masses are listed in the second column of the initial snapshot. Thus, we plot a histogram of the slice `data[0,:,1]` using the `hist()` function of pyplot. With the optional argument `histtype='step'`, the histogram is displayed as a step function (see Fig. 4.16). The mass distribution is nearly uniform, which is not quite the case in real stellar clusters (a uniform distribution is assumed in the simulation for simplicity). The mean value can be calculated with

```
26  print("Average mass: {:.3f} solar masses".
27          format(np.mean(data[0,:,1])))
```

and is close to the center of the mass $[0, 10\,M_\odot]$:

```
Average mass: 4.984 solar masses
```

Since the stars do not exchange mass, the mass distribution is constant in time. But the positions and velocities evolve under the action of gravity. How does the structure of the cluster change over time? To answer this questions, we compute the radial distances of all stars from the center and the and velocity magnitudes. The x, y, and z coordinates are stored in columns 3, 4, and 5 (array indices 2, 3, and 4) and the velocity components in the last three columns:

```
28  from astropy.constants import au,pc
29
30  r = np.sqrt(data[:,:,2]**2 +
31              data[:,:,3]**2 +
32              data[:,:,4]**2) * au/pc
33
34  v = np.sqrt(data[:,:,5]**2 + data[:,:,6]**2 + data[:,:,7]**2)
```

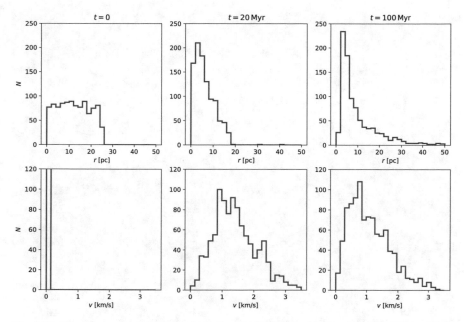

Fig. 4.17 Histograms of radial distances (blue) and velocities (red) of stars for three different times

In the following, we use pc as unit for radial distances, which is comparable to the scale of star clusters. Since the N-body code that produced the data uses AU as length unit, we multiply all radial distances by the conversion factor au/pc in line 32.

Now we can easily produce histograms for different snapshots to study the evolution of the cluster. For example, the following code combines three histograms for the radial distances at $t = 0$, 20, and 100 Myr and the corresponding velocity histograms in a multiple plot (see Fig. 4.17).

```
35  plt.figure(figsize=(12,8), dpi=200)
36
37  n_bins = 25
38
39  plt.subplot(231)
40  plt.hist(r[0,:], n_bins, range=[0,50],
41          histtype='step', lw=2, color='mediumblue')
42  plt.xlabel("$r$ [pc]")
43  plt.ylabel("$N$")
44  plt.ylim(0,250)
45  plt.title("$t=0$")
46
47  plt.subplot(232)
48  plt.hist(r[100,:], n_bins, range=[0,50],
49          histtype='step', lw=2, color='mediumblue')
```

```
50  plt.xlabel("$r$ [pc]")
51  plt.ylim(0,250)
52  plt.title("$t={:.0f}\,$Myr".format(100*0.2))
53
54  plt.subplot(233)
55  plt.hist(r[500,:], n_bins, range=[0,50],
56          histtype='step', lw=2, color='mediumblue')
57  plt.xlabel("$r$ [pc]")
58  plt.ylim(0,250)
59  plt.title("$t={:.0f}\,$Myr".format(500*0.2))
60
61  plt.subplot(234)
62  plt.hist(v[0,:], n_bins, range=[0,3.5],
63          histtype='step', lw=2, color='red')
64  plt.xlabel("$v$ [km/s]")
65  plt.ylabel("$N$")
66  plt.ylim(0,120)
67
68  plt.subplot(235)
69  plt.hist(v[100,:], n_bins, range=[0,3.5],
70          histtype='step', lw=2, color='red')
71  plt.xlabel("$v$ [km/s]")
72  plt.ylim(0,120)
73
74  plt.subplot(236)
75  plt.hist(v[500,:], n_bins, range=[0,3.5],
76          histtype='step', lw=2, color='red')
77  plt.xlabel("$v$ [km/s]")
78  plt.ylim(0,120)
79
80  plt.savefig("cluster_hist_evol.pdf")
```

We use subplot() to place plots radial and velocity histograms in the first and second row, respectively (see also Sect. 4.2). The three columns show the time evolution of the cluster from its initial to the final configuration as indicated by the labels on top of the plots. As we can see in Fig. 4.17, the initial distribution of stars is approximately uniform in the radial bins for $r \leq 25$ pc and all stars have small initial velocity. As time goes by, we see a concentration within the central 10 pc, while a sparse population emerges farther out. The velocity distribution also shifts and broadens, with a peak at roughly 1 km/s and just a few objects beyond 3 km/s. Since the cluster is a closed system that can neither gain nor lose energy in total, gravitational potential energy must have been transformed into kinetic energy in the process of stars concentrating near the center until an equilibrium is reached.

To examine the evolution in more detail, we compute averaged quantities for all stars at a given time. In the following, we consider root mean square (RMS) radial distances and velocities, which are defined by

$$r_{\mathrm{RMS}} = \langle x^2 + y^2 + z^2 \rangle^{1/2}, \qquad (4.76)$$

$$v_{\mathrm{RMS}} = \langle v_x^2 + v_y^2 + v_z^2 \rangle^{1/2}. \qquad (4.77)$$

The brackets $\langle\,\rangle$ denote averages over all stars. The RMS velocity is related to mean kinetic energy, assuming that all stars have the same mass (in Exercise 4.17 you are asked to compute the mean kinetic energy exactly):

$$\frac{E_{\mathrm{kin}}}{N} \sim \frac{1}{2} m v_{\mathrm{RMS}}^{1/2}.$$

where E_{kin} is the total kinetic energy of N stars. The computation of RMS values is straightforward with NumPy:

```
81  r_rms = np.sqrt(np.mean(r**2, axis=1))
82  v_rms = np.sqrt(np.mean(v**2, axis=1))
```

The function mean() with the keyword argument axis=1 computes mean values across columns (axis 1 refers to the column direction, while axis 0 is the row direction). Since the rows of r and v contain the data for subsequent snapshots, the calls of np.mean() in lines 81–82 return arrays of size given by the number of snapshots. By taking the square root, we get RMS values. For the radial distances, we also compute the medians, i.e. the value of r for which half of the stars are located at smaller distances and the other half at larger distance:

```
83  r_median = np.median(r, axis=1)
```

The results are readily plotted as functions of time:

```
84  t = np.linspace(0, 100, n_files)
85
86  plt.figure(figsize=(10.5,3.5), dpi=200)
87
88  plt.subplot(121)
89  plt.plot(t, r_rms, color='mediumblue')
90  plt.plot(t, r_median, ls='dashed', color='mediumblue')
91  plt.xlabel("$t$ [Myr]")
92  plt.ylabel("$r_\mathrm{RMS}$ [pc]")
93  plt.ylim(0,30)
94
95  plt.subplot(122)
96  plt.plot(t, v_rms, color="red")
97  plt.xlabel("$t$ [Myr]")
98  plt.ylabel("$v_\mathrm{RMS}$ [km/s]")
99  plt.ylim(0,1.8)
00
01  plt.savefig("cluster_evol_rms.pdf")
```

In Fig. 4.18 we see that the cluster contracts for about 25 million years until v_{RMS} reaches a maximum and bounces back. Around $t = 40$ Myr, the cluster seems to have reached an equilibrium with $v_{RMS} \approx 1.3$ km/s (this behaviour is similar to a damped oscillation). Interestingly, r_{RMS} keeps growing, suggesting that the size of the cluster gradually increases. However, outliers have a relatively strong impact on RMS values. The increase of r_{RMS} in time does not necessarily mean that the bulk of stars tends to move at larger distances from the center. Indeed, the median of r implies that the probability of finding stars within 7 pc stays at about 50% at late times. Even so, the cluster is only in quasi equilibrium (also keep in mind that neither r_{RMS} nor the median are directly related to the potential energy $\sim 1/r$). The steady growth of r_{RMS} is caused by evaporation, i.e. once in a while a star gets kicked out of the core region and can even escape the potential well of the cluster (see following exercises).

Exercises

4.16 Compare the median and RMS values of r for the cluster data used in this section to the maximum and the 90-th percentile (i.e., the value of r below which 90 % are found). Also prepare scatter plots of x- and y coordinates of all stars in pc in steps of 20 Myr (use pyplot's `scatter()` function). Does your analysis support the notion of cluster evaporation?

4.17 Compute the total kinetic and potential energy of the cluster for representative snapshots ranging from $t = 0$ to 100 Myr. For N mass points,

$$E_{kin} = \sum_{i=1}^{N} \frac{1}{2} m_i v_i^2, \qquad E_{pot} = -\sum_{i=2}^{N} \sum_{j<i} \frac{G m_i m_j}{r_{ij}}, \qquad (4.78)$$

where $r_{ij} = |\mathbf{r}_i - \mathbf{r}_j|$ is the distance between the mass points with indices i and j. Compare your results to the relation $2E_{kin} = -E_{pot}$ for a system in virial equilibrium.

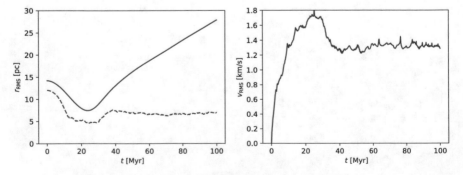

Fig. 4.18 Temporal evolution of the stellar RMS (solid line) and median (dashed line) distances to the coordinate center and the RMS velocity

Does the cluster relax toward equilibrium in time? How is the deviation from virial equilibrium affected if you exclude all stars beyond a certain percentile of r. Based on your findings, what would you suggest as a definition of the cluster core radius?

4.18 Extend the module nbody from Appendix A by a symplectic N-body solver using direct force summation. Start with a low number of particles, such as 10, and set random initial positions and velocities (similar to the cluster discussed in this section). Measure the performance of your solver using the %timeit tool (see Appendix B.1). Carefully increase the particle number. How far can you go before the computation slows down too much and a single integration step takes minutes or even longer?

4.6 Expansion of the Universe

At the end of this chapter, we return to first-order differential equations. We will solve an equation that describes the cosmological expansion of the Universe. This will allow us to determine the Hubble parameter $H(t)$, which is defined by ([4], Chap. 29)

$$H(t) = \frac{\dot{a}}{a} \tag{4.79}$$

where a is the time-dependent scale factor of the Universe and $\dot{a}$ its time derivative. The scale factor basically describes how the mean distance between galaxies changes due the expansion of the Universe, assuming a constant population of galaxies in a homogeneous and isotropic Universe. It is normalized to unity at present time. As we look back in cosmic history, the scale factor becomes smaller and smaller and vanishes at the Big Bang. For example, a scale factor $a = 0.5$ at some earlier time indicates that galaxies were typically closer by a factor of two at that time.

The expansion rate of the Universe is not constant.[27] The time evolution of the scale factor is governed by the Friedmann equation[28]:

$$\dot{a} = H_0 \sqrt{\frac{\Omega_{M,0}}{a} + \frac{\Omega_{rad,0}}{a^2} + \Omega_{\Lambda,0} a^2 + 1 - \Omega_0}, \tag{4.80}$$

where H_0 is the current value of the Hubble parameter and the parameters $\Omega_{M,0}$, $\Omega_{rad,0}$, and $\Omega_{\Lambda,0}$ measure the matter and energy content relative to the so-called critical density (this is the density for which the Universe just continues to expand

---

[27] As a result, the Hubble parameters changes in time. For this reason, the commonly used term Hubble constant is misleading. The Hubble parameter is only constant in space.

[28] While Einstein was convinced that the Universe must be static, the Russian physicist Alexander Friedmann derived the equation for the scale factor and found analytical solutions in the case $\Omega_{\Lambda,0} = 0$. His conclusion that the Universe is expanding was confirmed a few years later by Edwin Hubble.

forever and has a spatially flat geometry). More specifically, $\Omega_{M,0}$ is the total matter density, $\Omega_{rad,0}$, the radiation density, and $\Omega_{\Lambda,0}$ the density of dark energy (also called cosmological constant Λ). The subscript 0 indicates that the densities are measured at current time. The density parameters are dimensionless quantities. Cosmological observations indicate the total density is very close to the critical density, i.e.

$$\Omega_0 = \Omega_{M,0} + \Omega_{rad,0} + \Omega_{\Lambda,0} \simeq 1 .$$

Measurements of the cosmic microwave background (CMB) by the Planck spacecraft imply $\Omega_{M,0} \approx 0.309$, $\Omega_{\Lambda,0} \approx 0.691$, and $H_0 \approx 67.7$ km s^{-1}Mpc^{-1} [15].[29] This is the foundation of the so-called ΛCDM model (CDM is the abbreviation for cold dark matter), which is the currently accepted standard model of cosmology.

In the following example, the Hubble constant is defined with the help of Astropy units. We collect the cosmological parameters for different scenarios in a dictionary. In addition to the ΛCDM model with the keyword `'standard'`, we have matter-dominated models with $\Omega_0 < 1$, $\Omega_0 = 1$, and $\Omega_0 > 1$:

```
1  import numpy as np
2  import astropy.units as unit
3  from numkit import rk4_step
4
5  # Hubble constant with astropy units
6  H0 = 67.7*unit.km/unit.s/unit.Mpc
7  print("H0 = {:.2e}".format(H0.to(1/unit.Gyr)))
8
9  # dictionary of cosmological models
10 cosmology = {
11     'standard'    : (0.309, 1e-5, 0.691),
12     'matter sub'  : (0.309, 0, 0),
13     'matter crit' : (1, 0, 0),
14     'matter super': (2, 0, 0),
15 }
```

Expressed in units of 1/Gyr, the Hubble constant (i.e. the current value of the Hubble parameter) is

```
    H0 = 6.92e-02 1 / Gyr
```

The Friedmann equation is a first-order differential equation of the form $\dot{a} = f(t, a)$, where $f(t, a)$ is defined by the right-hand side of Eq. (4.80):

---

[29]The unit of H_0 originates from the Hubble law, which relates the distance (Mpc) and recession speed (km s^{-1}) of an object due to the cosmological expansion. Other measurements of H_0, e.g. based on distant supernovae, indicate a somewhat higher value above 70 km s^{-1}Mpc^{-1}. The cause of this discrepancy is not fully understood yet.

```
16  def dota(t, a, OmegaM, OmegaR, OmegaL, H0):
17      Omega0 = OmegaM + OmegaR + OmegaL
18      return H0 * (OmegaM/a + OmegaR/a**2 + OmegaL*a**2 +
19                   1 - Omega0)**(1/2)
```

As preparation for the numerical integration of the Friedmann equation, we check the convergence of the value of the scale factor with decreasing time step. The characteristic time scale is the Hubble time

$$t_{\mathrm{H}} = 1/H_0 \,. \tag{4.81}$$

This is roughly the time after which the Universe has expanded to the current scale factor $a(t_0) = 1$, i.e. $t_{\mathrm{H}} \sim t_0 \sim 10$ Gyr. For our convergence test, we integrate the standard model from $t = 0$ to $t_{\mathrm{max}} = 0.1 t_{\mathrm{H}}$ and compare the final values of the scale factor obtained with different time steps.

```
20  # numerical values (time in units of Gyr)
21  H0_num = H0.to(1/unit.Gyr).value
22  t_H = 1/H0_num
23
24  t_max = 0.1*t_H
25  n = 10
26
27  while n <= 1000:
28      t, a = 0, 0.01  # initial values
29      dt = t_max/n    # time step
30
31      # numerical integration from 0 to t_max
32      for i in range(n):
33          a = rk4_step(dota, t, a, dt,
34                       *cosmology['standard'], H0_num)
35          t += dt
36
37      print("{:4d} {:.8e}".format(n,a))
38      n *= 2
```

We express time in units of Gyr and start with a time step of $0.1 t_{\mathrm{max}}$ (10 time steps). The Friedmann equation is solved by means of the Runge-Kutta method introduced in Sect. 4.1.1. Here, the implementation from numkit is applied. The density parameters from the dictionary and the Hubble parameter (in units of 1/Gyr) are passed as variadic arguments via the argument list of rk4_step() to the function dota() for the calculation of the time derivative of the scale factor (see lines 33–34). The tuple cosmology['standard'] is split by the unpacking operator *, which is equivalent to the call

```
rk4_step(dota, t, a, dt,
          cosmology['standard'][0], cosmology['standard'][1],
          cosmology['standard'][2], H0_num)
```

The number of time steps is doubled after each iteration of the outer loop (see line
38) and terminates once the number exceeds 1000. The resulting scale factors at time
t_max are:

```
 10 1.93079317e-01
 20 1.92796594e-01
 40 1.92753019e-01
 80 1.92748144e-01
160 1.92747742e-01
320 1.92747714e-01
640 1.92747713e-01
```

Since RK4 is a higher-order method, the result converges fast (the relative precision is $\sim 10^{-8}$ for 640 time steps). How far should we go? Since the uncertainty of
the cosmological parameters is about 10^{-3}, it is reasonable to reach a comparable
precision. Our test suggests that about 50 time steps will be sufficient.

Now we proceed to compute the evolution of the scale factor for all models from
$t = 0$ to $2t_{\mathrm{H}}$. Since this time interval is 20 times longer, we need 1000 time steps.
For each set of parameters, the Friedmann equation is numerically integrated and the
resulting data for the scale factor are plotted:

```
42  import matplotlib.pyplot as plt
43  %matplotlib inline
44
45  fig = plt.figure(figsize=(6,4), dpi=100)
46
47  n = 1000
48  dt = 2*t_H/n
49  t = np.linspace(0, 2*t_H, n+1)
50
51  for model in cosmology:
52      a = np.zeros(n+1)
53      a[0] = 1e-2
54
55      # numerical integration of the model
56      for i in range(n):
57          a[i+1] = rk4_step(dota, t[i], a[i], dt,
58                            *cosmology[model], H0_num)
59
60      # plot the scale factor as function of time
61      label = "$\Omega_{\mathrm{M}}=$"
62      label += "{:.1f}, $\Omega_\Lambda=${:.1f}".\
```

```
63            format(cosmology[model][0],cosmology[model][2])
64        if model == "standard":
65            plt.plot(t, a, label=label)
66        else:
67            plt.plot(t, a, ls='dashed', label=label)
68
69  plt.xlabel("$t$ / Gyr")
70  plt.ylabel("$a$")
71  plt.legend()
72  plt.savefig("scale_fct_evol.pdf")
```

In Fig. 4.19, the ΛCDM model is shown as solid line and the matter-dominated models as dashed lines. The main difference between the models is that dark energy causes an accelerated expansion when the term $\Omega_{\Lambda,0}a^2$ under the square root in Eq. (4.80) dominates. In the other cases, the expansion decelerates. If the total density parameter, Ω_0, is below unity, the universe will keep expanding forever, but at an ever decreasing rate. For $\Omega_0 < 1$, on the other hand, the scale factor reaches a maximum and then the universe begins to collapse until it finally ends in a Big Crunch.

In the presence of dark energy, there is a turning point where the slope $\dot{a}$ switches from decreasing to increasing. The condition for this is $\ddot{a} = 0$. At which time did the transition occur in our Universe? To answer this question, we need to find the time when $\ddot{a}$ vanishes. The simplest method is to compute the second derivate from our data for the scale factor in a loop. Once the sign switches from negative (decelerating) to positive (accelerating), we have found the turning point:

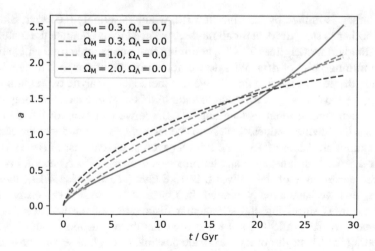

Fig. 4.19 Evolution of the cosmological scale factor for models with different matter and dark energy density parameters. The solid line corresponds to our Universe (based on data from the Planck mission)

```
73 │ n = int(t_H/dt)
74 │ a = np.zeros(n+1)
75 │ a[0] = 1e-2
76 │
77 │ for i in range(n):
78 │     a[i+1] = rk4_step(dota, i*dt, a[i], dt,
79 │                       *cosmology['standard'], H0_num)
80 │
81 │ # compute second derivative of scale factor
82 │ # and terminate if sign reverses
83 │ i = 0
84 │ ddota = -1 # arbitrary start value
85 │ while ddota < 0 and i < n:
86 │     i += 1
87 │     # second-order centered differences
88 │     ddota = (a[i+1] - 2*a[i] + a[i-1])/dt**2
89 │
90 │ if ddota >= 0:
91 │     print("Transition time = {:.2f} Gyr".format(i*dt))
92 │ else:
93 │     print("Transition time not reached")
```

The answer is

```
Transition time = 7.63 Gyr
```

The accuracy is limited by the time step dt, which is roughly 0.01 Gyr. Since we did not bother to store the data for all models, we need to re-compute the scale factor for the standard model (lines 77–79), assuming that one Hubble time will suffice to find the turning point of $a(t)$. Why do we not apply one of our root finders here to determine the zero of the function $\ddot{a}$? Well, if there is no analytic expression for the function, we have to work with discrete data. In this case, the data are computed by means of centered differences (see line 88). Since we can monitor the sign while computing the function values, there is no need to apply a root finder to the complete array of function values afterwards. In other words, we can test in a **while** loop whether ddota is still negative and terminate as soon as ddota becomes positive. Since the current age of the Universe is 13.8 Gyr (see Exercise 4.19), the result indicates that we have already entered the phase of accelerated expansion. It was one of the most surprising and spectacular discoveries of the last decades when astronomers realized in 1998 for the first time that the expansion of our Universe is accelerating [16]. This discovery was made possible through observations of distant supernovae.

Before the role of dark energy was recognized, it appeared more likely that our Universe should be matter dominated. For comparison, Fig. 4.19 shows the expansion history of an alternative universe with the same matter content, but no dark energy (radiation is also neglected, but this makes a difference only in the very early phases with $a \ll 1$). In this case, the expansion rate becomes asymptotically constant:

$$\dot{a} \simeq H_0 \sqrt{1 - \Omega_0} \quad \text{for } a \to \infty.$$

For $\Omega_0 = \Omega_{M,0} = 0.3$, the asymptotic value is $\approx 0.55 H_0$. If the total density is critical ($\Omega_0 = 1$), then $\dot{a}$ will stagnate at zero. For even higher matter density, the expansion reverses at a maximum scale factor

$$a_{max} = \frac{\Omega_{M,0}}{\Omega_0 - 1}.$$

This formula yields $a_{max} = 2$ for the model with the keyword `"matter super"`.

How long does it take such a universe to reach a_{max}? The answer is obtained by integrating the differential equation $dt = da/\dot{a}$, where $\dot{a}$ is given by Eq. (4.80):

$$t_{max} = \int_0^{a_{max}} \frac{da}{\dot{a}}. \tag{4.82}$$

The time of the Big Crunch is $2t_{max}$ because expansion and collapse are symmetric with respect to $t = t_{max}$. For arbitrary density parameters, the integral can be solved numerically. The following code utilizes the function `integrate.quad()` from SciPy,[30] which can handle the divergent expansion rate at $a = 0$. In Exercise 4.19, you are asked to solve the integral alternatively with our implementation of Simpson's rule. Since `integrate.quad()` does not support variadic arguments, we use a Python lambda with fixed density parameters:

```
94  from scipy import integrate
95
96  OmegaM, OmegaR, OmegaL = cosmology['matter super']
97  Omega0 = OmegaM + OmegaR + OmegaL
98
99  tmp = integrate.quad(
00      lambda a: (OmegaM/a + OmegaR/a**2 + OmegaL*a**2 +
01                 1 - Omega0)**(-1/2),
02      0, 2)
03
04  t_crunch = 2*tmp[0]/H0_num
05  print("Time of Big Crunch: {:.1f} Gyr".format(t_crunch))
```

The matter-dominated model universe with twice the critical density needs 45 billion years to reach its maximum expansion and it finally perishes at

```
Time of Big Crunch: 90.7 Gyr
```

---

[30] See docs.scipy.org/doc/scipy/reference/integrate.html for more information about numerical integration with SciPy. The name quad comes from the term "quadrature" that is sometimes used for integration.

Exercises

4.19 Compute the current age $(a = 1)$ of our Universe from Eq. (4.82). How old would the Universe be if there was no cosmological constant? Use both `integrate.quad()` from `scipy` and `integr_simpson()` from `numkit` and compare the results.

4.20 Apply `integrate.solve_ivp()` (see Sect. 4.3) to solve the Friedmann equation. How large is the deviation from the solution computed with the Runge-Kutta integrator applied in this section? Can you reduce the deviation? Study the outcome for $\Omega_{\Lambda,0}$ smaller and greater than the value in our Universe.

4.21 Owing to the cosmological expansion, distant galaxies move away from us and appear red-shifted. The light we observe now was emitted in the past, when the Universe was younger and the scale factor smaller. As a consequence, the cosmological redshift z of a distant galaxy is related to the scale factor a by $a = 1/(1 + z)$. By substituting this relation into the Friedmann equation, we obtain an equation for the redshift-dependent Hubble parameter:

$$H(z) = H_0\sqrt{\Omega_{M,0}(1 + z)^3 + \Omega_{rel,0}(1 + z)^4 + \Omega_{\Lambda,0} + (1 - \Omega_0)(1 + z)^2}$$

This allows us to compute the distance d_0 of a galaxy with redshift z ([4], Sect. 29.4):

$$d_0 = c\int_0^z \frac{dz'}{H(z')}$$

where c is the speed of light. Compute d_0 for galaxies observed at $z = 1$ in the universes investigated in Exercise 4.20 (i.e. for different dark energy densities).

The distance d_0 is a so-called proper distance. For observations, astronomers use the luminosity distance of a galaxy. It is defined by the ratio of the luminosity of a galaxy to the observed radiative flux:

$$d_L^2 := \frac{L}{4\pi F}$$

Since cosmological expansion affects the propagation of light, it is related to proper distance by $d_L = (1 + z)d_0$.[31] Compute the luminosity distance as a function of redshift, $d_L(z)$, for 30 logarithmically spaced points in the interval $0.01 < z < 10$.

- Verify the observational Hubble law $d_L = cz/H_0 z$ for small redshifts $(z \ll 1)$.
- Plot the distance modulus (see Exercise 3.2)

---

[31] This follows from general relativity. Apart from redshifting (stretching of the wavelength), photons travelling from distant galaxies to Earth experience time dilatation, which in turn alters the energy received per unit time compared to smaller, non-cosmological distances.

$$m - M = 5 \log_{10} \left(\frac{d_{\mathrm{L}}}{10 \ \mathrm{pc}} \right)$$

as a function of z (logarithmic scale). This relation was used to deduce the accelerated expansion of our Universe from observations of supernovae of type Ia in distant galaxies [16]. At which redshifts do you see significant differences depending on the dark energy density $\Omega_{\Lambda,0}$? How large is the difference in magnitudes for $z = 1$?

Chapter 5
Astronomical Data Analysis

Abstract Astronomy and astrophysics are highly data-driven research fields: Hypotheses are built upon existing data, models are used to make predictions and discrepancies between theory and observation drive scientific progress, forcing us to either modify existing models or come up with new solutions. In this chapter, we discuss techniques for analysing a variety of data, ranging from individual stellar spectra and light curves to large surveys, such as GAIA. Naturally, file input and output are an important prerequisite for data processing. We conclude with a brief introduction to convolutional neural networks and their application to image data and spectra.

5.1 Spectral Analysis

In Chap. 3, we discussed how stars can be classified by spectral properties. Modern spectrographs can produce spectra with high wavelength resolution, which allows for the detailed analysis of absorption lines. As an example, the online material for this book (uhh.de/phy-hs-pybook) includes an optical spectrum of the star ζ *Persei* taken with the Ultraviolet and Visual Echelle Spectrograph (UVES) of ESO. The spectrum is stored in the FITS file format, which can be considered a de-facto standard for astronomical data. In contrast to plain text (ASCII) based formats, FITS files are binary, which reduces the operational overhead during read and write processes. Furthermore, every FITS file contains a header with a detailed description of the data and its format. This is information is called metadata.

Thanks to the Astropy library, FITS files can be accessed with just a few lines of code. Tools for handling FITS files are provided by the module `astropy.io`:

```
1  from astropy.io import fits
2  import matplotlib.pyplot as plt
3  import numpy as np
```

After specifying the path and name of the FITS file, we can load its contents with the `open()` function and display some basic properties:

© Springer Nature Switzerland AG 2021
W. Schmidt and M. Völschow, *Numerical Python in Astronomy and Astrophysics*,
Undergraduate Lecture Notes in Physics,
https://doi.org/10.1007/978-3-030-70347-9_5

```
4  file = "data_files/ADP.2014-10-29T09_42_08.747.fits"
5  fits_data = fits.open(file)
6  fits_data.info()
```

This code prompts the metadata of the file (last line abbreviated):

```
Filename: ADP.2014-10-29T09_42_08.747.fits
No.    Name     Ver    Type      Cards   Dimensions   Format
  0  PRIMARY      1  PrimaryHDU    788   ()
  1  SPECTRUM     1  BinTableHDU    71   1R x 6C   [134944D, 134944E, ..., 134944E]
```

In the table, the HDUs (Header Data Units) of the file are listed. The second HDU contains the full spectrum, which is tabulated in six columns. For a more detailed description of the columns, for example, the physical units of the quantities, we can execute the **print**(fits_data[1].columns) command, leading to the following output:

```
ColDefs(
    name = 'WAVE'; format = '134944D'; unit = 'Angstrom'
    name = 'FLUX_REDUCED'; format = '134944E'; unit = 'adu'
    name = 'ERR_REDUCED'; format = '134944E'; unit = 'adu'
    name = 'BGFLUX_REDUCED'; format = '134944E'; unit = 'adu'
    name = 'FLUX'; format = '134944E'; unit = '10^-16 erg/cm^2/s/Angstrom'
    name = 'ERR'; format = '134944E'; unit = '10^-16 erg/cm^2/s/Angstrom'
)
```

Specifically, we are interested in columns 0 and 4, i.e. the wavelength and flux. The entire spectrum is accessible via fits_data[1]. To work with the data, we first extract the full spectrum and dump it into a new array, after which the FITS file can be closed:

```
7  scidata = fits_data[1].data
8  fits_data.close()
```

It is important that scidata has a FITS-specific type that is inherited from numpy. You can check with **type**() and **isinstance**(), as described in Sect. 2.1.3. Formally, scidata is defined as an array with just one row and six columns. For the sake of easy referencing, we further extract the desired columns and copy them into one-dimensional NumPy arrays:

```
9   wavelength = scidata[0][0]
10  flux = scidata[0][4]
```

As we do not require absolute spectral flux values, we normalize the spectrum by dividing through the peak value. Moveover, we convert wavelengths from Angstrom to nanometers:

```
11  norm = np.max(flux)
12  flux = flux/norm
13  wavelength = wavelength*0.1
```

Now we are prepared to display the data. The code

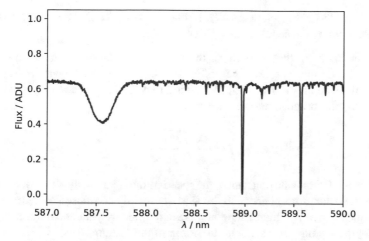

Fig. 5.1 Helium and sodium absorption lines in spectrum of ζ Persei

```
14   %matplotlib inline
15
16   plt.plot(wavelength, flux, linestyle='-' , color='navy')
17   plt.xlabel("$\lambda$ / nm")
18   plt.ylabel("Flux / ADU")
19   plt.xlim(587,590)
20
21   plt.savefig("spectrum_full.pdf")
```

produces Fig. 5.1. The plot shows the spectrum of ζ Persei in the wavelength range from 587 to 590 nm. One can see two very distinct types of absorption features. Since ζ Persei is a B-type supergiant with an effective temperature of 20 800 K, helium absorption lines such as the broad line at 587.6 nm can be seen. However, observers noticed already int the early 20th century that distant stars tend to show unexpected features in their spectra that are unlikely to originate from a stellar photosphere. This led to the discovery of *matter between the stars*, the so-called interstellar medium (ISM). An example are the two narrow lines at 589.0 and 589.6 nm, which originate from sodium in the ISM. These lines were discovered in spectroscopic binary stars by Mary L. Heger in 1919. She realised that their width was at odds with the broad appearance of other absorption lines in the stellar spectrum. You can further explore these lines and the spectrum of ζ Persei in the following exercises.

Exercises

5.1 Investigate the spectrum of ζ Persei in the wavelength range from 480 to 660 nm. Use plt.axvline() to mark the Balmer absorption lines Hα and Hβ at wavelengths 656.3 and 486.1 nm, respectively, by vertical dashed lines (see the Matplotlib online documentation and the next section for examples). You can label the lines with the help of plt.text(). In addition to the Balmer lines, try to identify the absorp-

tion lines of He I at wavelengths 471.3, 492.1, 501.6, 504.7, 587.6, and 667.8 nm
and mark them by dashed lines in a different color.

5.2 Devise an algorithm to estimate the full width at half-maximum, $\lambda_{1/2}$, of the
helium and sodium absorption lines in Fig. 5.1. $\lambda_{1/2}$ measures the width of a spectral
line at half its depth [4, Sect. 9.5]. Assuming that the line broadening is caused by
thermal Doppler broadening, we have

$$\lambda_{1/2} = \frac{2\lambda}{c} \sqrt{\frac{2kT \log 2}{m}},$$

where k is the Boltzmann constant, c the speed of light, and m the mass of the atoms.
Since both stellar atmospheres and the ISM are mainly composed of hydrogen, you
can assume $m \approx m_H$. Which temperatures are implied by your estimates of $\lambda_{1/2}$ and
what do your results suggest about the origin of the sodium lines?

5.2   Transit Light Curves

The last three decades have seen a dramatic revolution in our understanding of planetary systems and, thanks to ever-decreasing instrumental thresholds, the detection
of exoplanetary systems has become daily routine. As of fall 2020, about 4,300 confirmed exoplanets in more than 3,000 stellar systems have been found, including
planets around binary or even tertiary stars.[1] The most surprising discovery have
been so-called *Hot Jupiter* systems which contain at least one Jovian planet in close
proximity to a star. In the scatter diagram shown Fig. 5.2, hot Jupiters are found in
the left upper region (small-major axis and large mass; see also Exercise 2.9).

The majority of confirmed planetary systems has been detected via planetary
transits or periodic variations of a star's radial velocity. In the following, we will
focus on the transit method. When the orbital plane of an exoplanet happens to be
nearly aligned with the line of sight from Earth, the exoplanet passes between us
and its hosting star, causing it to block a fraction of the light emitted by the star and
reducing the measured flux (see Fig. 5.3).

Assuming a uniform surface brightness and given the stellar and planetary radius
R_S and R_P, respectively, the fraction of blocked light can be estimated via

$$\frac{\Delta F}{F} = \frac{4\pi R_P^2}{4\pi R_S^2} = \left(\frac{R_P}{R_S}\right)^2. \tag{5.1}$$

For a Jovian planet orbiting a solar-like star, we have $R_P/R_S \approx 0.1$ implying that
only 1% of the star's light will be blocked by the planet. Things get even worse for
an Earth-like planet where we have $R_P/R_S \approx 0.01$, i.e. only 0.01% blocked light.

---

[1]Resource for current data are exoplanets.nasa.gov and www.exoplanet.eu.

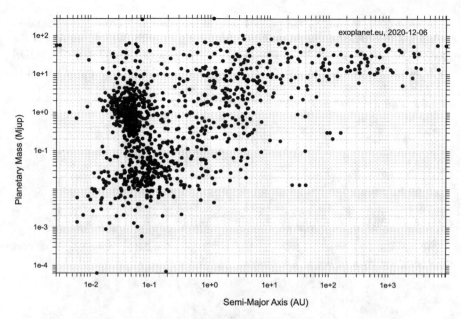

Fig. 5.2 Distribution of (confirmed) exoplanet masses as a function of the orbit's semi-major axis (Diagram generated with interactive website www.exoplanet.eu/diagrams)

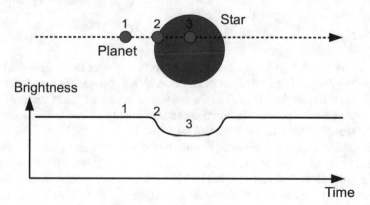

Fig. 5.3 Illustration of an exoplanet transit (Image credit: Hans Deeg, commons.wikimedia.org)

While the detection of terrestrial planets requires both large telescopes and exceptional atmospheric conditions, which are only achieved at today's top astronomical facilities, hot Jupiter systems are routinely detected by smaller telescopes with limited seeing.

In the online resources, we provide a light curve of the TrES-2 system taken by the 1.2 m *Oskar-Lühning-Teleskop* in Hamburg in June 2013.[2] The ASCII file `tres2_data.dat` contains three columns, namely the modified Julian date MJD,[3] the relative flux and the flux error. A light curve is the time-dependent record of the incident flux received from an object. First, we'll load the data using NumPy's `loadtxt()` function and then slice the columns into separate arrays:

```
1  import numpy as np
2
3  data = np.loadtxt("tres2_data.dat")
4
5  mjd = data[:,0]
6  flux = data[:,1]
7  err = data[:,2]
```

We can use `errorbar()` from `pyplot` to plot the data points with error bars indicating the error of the measurement:

```
8   import matplotlib.pyplot as plt
9   %matplotlib inline
10
11  plt.errorbar(mjd, flux, yerr=err, ecolor='steelblue',
12               linestyle='none', marker='o', color='navy')
13  plt.xlabel("MJD")
14  plt.ylabel("Flux / ADU")
15
16  plt.savefig("tres2_lightcurve.pdf")
```

This gives us Fig. 5.4. We can clearly see a dip in the light curve that spans about 100 min (the Julian date is in units of days), which is caused by the object known as TrES-2b or Kepler-1b.[4] However, the refractive nature of Earth's atmosphere in combination with temperature fluctuations and turbulent motions causes a significant jitter and scatter in the data we have to deal with.

For scientific analysis, one would fit complex transit models to the data. Here, we will identify basic properties by means of visual inspection and elementary calculations. First of all, we estimate the beginning of the transit, when the exoplanet just begins to move over the edge of the stellar disk (ingress), and the end, when it moves outside of the disk (egress)[5]:

---

[2]Named after Oskar Lühning, who was killed as a young man in World War II before he could see his wish to study astrophysics come true. The telescope was funded by a donation of Oskar's farther in memory of his son.

[3]The Julian date with the first two digits removed; see also Sect. 3.3.

[4]It was originally discovered in 2006. The naming convention for exoplanets is to add a lowercase letter to the name of the star system, which is often derived from an observational campaign. The letter 'b' indicates the first exoplanet detected in the system.

[5]The time interval $[T_1, T_4]$ covers the full length of the transit, while $[T_2, T_3]$ is the period when the exoplanet is entirely inside the stellar disk.

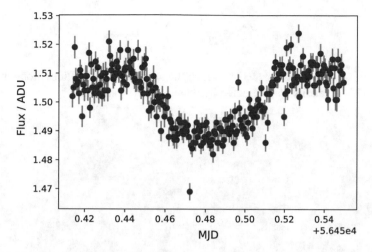

Fig. 5.4 Light curve of a transit of TrES-2b observed from Hamburg on July 6th, 2013

```
17   T1 = 5.645e4 + 0.445
18   T4 = 5.645e4 + 0.520
```

We can use these times to calculate a normalization factor from all flux values *outside* of the transit. NumPy allows us to perform this task effortlessly by selecting array values from flux based on a condition such as mjd<T1 (conditional indexing):

```
19   norm1 = np.mean(flux[mjd<T1]) # before transit
20   norm2 = np.mean(flux[mjd>T4]) # after transit
21   norm = 0.5*(norm1+norm2)
22
23   print(f"Flux normalization factor: {norm:.3f}")
24
25   # normalize fluxes
26   flux /= norm
27   err /= norm
```

In the last two lines, both the flux and the flux error are normalized by

```
Flux normalization factor: 1.509
```

To determine the transit depth, which in turn allows us to calculate the size of the exoplanet, we need to find a statistical value of the minimum flux, despite the fluctuations in the measurements. A relatively simple method to smooth the light curve is a moving average. Let us denote the ith measurement of the flux in the time series by F_i. We can smooth the flux by averaging over a given number, N, of neighbouring data points:

$$F_i^{(n)} = \frac{1}{N+1} \sum_{n=-N/2}^{N/2} F_{i+n} \tag{5.2}$$

With increasing i, the window over which the average is taken slides along the light curve.[6] This is readily implemented by means of array slicing:

```
28  # width and offset of sample window
29  offset = 7
30  width = 2*offset + 1
31
32  # compute moving average
33  flux_smoothed = np.ones(flux.size - width + 1)
34  for i,val in enumerate(flux_smoothed):
35      flux_smoothed[i] = np.sum(flux[i:i+width])/width
36
37  flux_min = np.min(flux_smoothed)
38  print(f"Minimum flux: {flux_min:.3f}")
```

The variables offset and width correspond to $N/2$ and $N+1$ in Eq. (5.2). In our example, we have $N = 15$, i.e. each smoothed value is an average over 15 data points. The minimum of the smoothed flux will be used below:

```
Minimum flux: 0.985
```

Let us plot now the smoothed light curve on top of the data point:

```
39  plt.errorbar(mjd, flux, yerr=err, ecolor='steelblue',
40               linestyle='none', marker='o', color='navy',
41               zorder=1)
42  plt.xlim(np.min(mjd), np.max(mjd))
43  plt.xlabel("MJD")
44  plt.ylabel("rel. flux")
45
46  # smoothed flux
47  plt.plot(mjd[offset:-offset], flux_smoothed,
48           lw=2, color='orange', zorder=2)
49
50  # ingress, egress, and minimum flux
51  plt.axvline(T1, color='crimson', lw=1, linestyle=':')
52  plt.axvline(T4, color='crimson', lw=1, linestyle=':')
53  plt.axhline(flux_min, lw=1, linestyle='--', color='black')
54
55  plt.savefig("tres2_lightcurve_smooth.pdf")
```

The outcome can be seen in Fig. 5.5. The array flux_smoothed is plotted in line 47. We have to take into account that this array has less elements and its index has an

---

[6]This is a variant of the simple moving average (SMA), which calculates an average from N previous measurements. In general, moving averages are defined by convolution over a window function of prescribed width and shape.

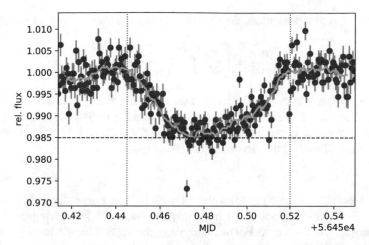

Fig. 5.5 Smoothed light curve computed from the data points plotted in Fig. 5.4. The vertical dotted lines indicate the duration of the transit and the horizontal dashed line the maximal reduction of the star's flux

offset by $N/2$ compared to the original data array. For this reason, we need to take the slice `mjd[offset:-offset]` from the time array. The keyword argument `zorder` in lines 41 and 48 brings the orange curve on top of the data points with error bars. In lines 51–52, the beginning of the ingress, `T1`, and the end of the egress, `T4`, are marked by vertical lines. Moreover, the minimum flux calculated above is shown as horizontal dashed line. Our result, $\Delta F/F \approx 1 - 0.985 = 0.015$, allows us to calculate the star-planet radius ratio. From Eq. (5.1), it follows that $R_P/R_S \approx 0.12$, which is typical for a Jovian planet orbiting a sun-like star.

Since the duration of the transit, $T_{trans} = T_4 - T_1$, depends on the orbital velocity and distance of the exoplanet from the star, it is possible to derive the size of the exoplanet through geometric reasoning and Kepler's third law. Assuming that the orbital plane is exactly aligned with the line sight, i.e., the exoplanet transits the center of the stellar disk, it follows that

$$\sin\left(\frac{\pi\, T_{trans}}{P}\right) = \frac{R_S + R_P}{a}$$

where P is the orbital period and a the semi-major axis of the exoplanet. The orbital period is also observable with the transit method: It is just the time between two subsequent transits ($P = 2.47063$ d for TrES-2b). Since T_{trans}/P is relatively small, we can apply the small-angle approximation. By rewriting the above equation in terms of the flux deficiency $\Delta F/F$, the following approximate relation for the planet radius is obtained:

$$R_P \simeq a\, \frac{\pi\, T_{trans}}{P}\left(1 + \sqrt{\frac{F}{\Delta F}}\right)^{-1} \tag{5.3}$$

Neglecting the planetary mass, the semi-major axis a is related to the orbital period P by

$$\frac{a}{1 \text{ AU}} = \left(\frac{M_S}{M_\odot}\right)^{1/3} \left(\frac{P}{365.25 \text{ d}}\right)^{2/3} \tag{5.4}$$

The mass of the star can be determined independently: $M_S = 0.98 \, M_\odot$, implying $a = 0.0355$ AU. Substitution of all parameters into Eq. (5.3) yields $R_P \approx 0.8 \, R_{\text{Jup}}$. By using a more accurate model (see Exercise 5.4), $R_P \approx 1.27 \, R_{\text{Jup}}$ is obtained. Thus, we can classify TrES-2b as a typical hot Jupiter.

Exercises

5.3 The estimate of $\Delta F/F$ from the smoothed light curve depends on several choices. Apart from T_1 and T_4, it is mainly the width of the moving average, N, that determines the outcome. For our analysis of the TrES-2 transit, these parameters can be regarded as *tuning* parameters. For a more systematic analysis, write a Python function that performs the computation form code line 19 to 37 for given start and end times of the transit and window width. The function should return the smoothed flux and its minimum.

(a) Vary the tuning parameters within reasonable bounds and analyse the sensitivity of $\Delta F/F$. Apply the law of error propagation to estimate how this affects the planet radius following from Eq. (5.3).
(b) Figure 5.5 suggests a relatively strong impact of outliers such as the very low flux value close to the halftime of the transition. Can you think of a criterion to exclude extreme outliers from the data? Check whether this reduces the sensitivity on tuning parameters.

5.4 If the orbital plane is not exactly aligned with the line of sight, the transit model becomes more complicated because T_{trans} also depends on the inclination i of the system. In the general case, it can be shown that

$$R_P = a \sqrt{1 - \sin i \, \cos\left(\frac{2\pi \, T_{\text{trans}}}{P}\right)} \left(1 + \sqrt{\frac{F}{\Delta F}}\right)^{-1} \tag{5.5}$$

The inclination can be obtained from a detailed analysis of the ingress and egress phases of the transit. In the case of the TrES-2 system, it was determined to be $i = 83.6°$. Apply the improved model in combination with the transit depth resulting from Exercise 5.3 to compute R_P.

5.3   Survey Data Sets

In the last decades, huge surveys of astronomical objects have been carried out, particularly by space telescopes. For example, the most exhaustive census of the galactic solar neighbourhood is made by the GAIA mission.[7] Basic properties of more than a billion stars have been accurately measured, representing roughly 1% of all stars within our galaxy. Most notably, GAIA measured stellar parallaxes, providing insight into the three-dimensional structure of the Milky Way.

The survey data can be accessed online via the GAIA archive.[8] In the archive's search menu, queries can be made by submitting ADQL commands under the tab *Advanced (ADQL)*. ADQL extends the SQL database language by various numerical commands and routines that are commonly used in astronomy. A typical query consists of three lines *selecting* data columns *from* a particular catalog and extracting data items *where* certain Boolean conditions are met:

```
SELECT l, b, parallax, parallax_over_error, radial_velocity, phot_g_mean_mag
FROM gaiadr2.gaia_source
WHERE phot_g_mean_mag<12 AND ABS(radial_velocity)>0 AND parallax>=1.0 AND parallax_over_error>=10
```

In this query, we choose the individual source catalogue of the *GAIA Data Release 2*. From the catalog, we select the galactic longitude l and latitude b, the parallax and its relative error, the radial velocity along the line of sight, and the mean G band magnitude as stellar properties. The meaning of galactic coordinates is explained in Fig. 5.6. In the last line, we apply several filters. First, we limit the number of objects by considering only stars brighter than 12th magnitude. Moreover, we only allow for objects with valid radial velocity values, parallaxes ≥ 1 mas (corresponding to distances smaller than 1 kpc) and relative parallax errors smaller than 10% in our sample. After processing our query the system reports that a total of 1,386,484 objects have been selected from the catalogue. You can submit this query yourself and download the resulting sample. We recommend using the CSV format (you can choose via the pulldown menu *download format*). The size is about 130 MB, so this might take a little while.

Once you have downloaded the data file, you can load into a NumPy array:

```
1  import numpy as np
2  import matplotlib.pyplot as plt
3
4  data = np.loadtxt("gaia_12mag_1kpc-result.csv",
5                    dtype='float64', usecols=(0, 1, 2, 4),
6
7                    delimiter=',', skiprows=1)
```

In the following, we will only make use of the galactic longitude and latitude, the parallax, and the radial velocity. For this reason, only columns 0, 1, 2, and 4 are used. Since the data file starts with a comment line with a leading # character,[9] which

---

[7]Mission website: sci.esa.int/web/gaia.

[8]Archive website: gea.esac.esa.int/archive.

[9]On Linux or Mac computers, you can easily check this on the shell with the help of the head command.

Fig. 5.6 Illustration of the galactic coordinate system. The galactic longitude *l* is the angle between the line of sight to a distant object projected into the galactic midplane (solid line) and the reference direction from the Sun to the centre of the Milky Way (dashed line). The latitude *b* is the angular distance measured from the galactic midplane

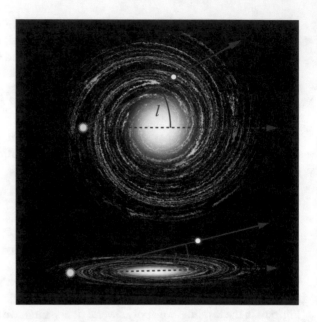

cannot be interpreted as a row of float numbers by `np.loadtext()`, we need to skip the first row. This is indicated by the keyword argument `skiprows`, which specifies the number of rows to be skipped (not to be confused with the index of the first row, which is zero). Moreover, `np.loadtext()` expects whitespaces as delimiters between columns by default. In the CVS table, however, comas are used. This needs to be specified by `delimiter=','`.

What are the dimensions of the array `data`?

```
8 | print(data.shape)
```

prints

```
(1386484, 3)
```

Even though we applied several filters, the data set is still fairly large with roughly 1.4 million rows and 3 columns. Histograms are a simple means of displaying statistical properties of such data samples. For a start, let us plot the distribution of stellar distances. The distance can be directly computed from the measured parallax π:

$$\frac{d}{1\text{ pc}} = \frac{1''}{\pi} \tag{5.6}$$

Since we have parallaxes in microarcseconds (mas), we get the distance in units of kpc by inverting the parallaxes:

```
9  | d = 1/data[:,2]
10 |
```

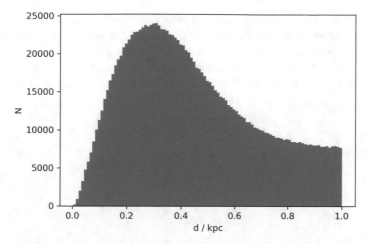

Fig. 5.7 Distribution of distances in a sample of 1.4 million stars from the GAIA archive

```
11  fig = plt.figure(figsize=(6, 4), dpi=300)
12
13  plt.hist(d, 100)
14  plt.xlabel('d / kpc')
15  plt.ylabel('N')
16  plt.savefig('d_histogram.png')
```

The histogram is shown in Fig. 5.7. For 100 bins (see line 12), the bin size is 1 kpc/100 = 10 pc. In the close neighbourhood of the Sun, the number of stars per distance bin increases up to a maximum located roughly at 0.3 kpc. Beyond that, the number of stars per bin decreases with distance. The histogram is shaped by several factors:

- Assuming a nearly constant star density, the number of stars dN in a spherical shell of thickness dr around the Sun scales with distance as $dN \propto r^2\, dr$. This determines the growth of N for small distances.
- Since our galaxy has a disk-like geometry with an average thickness of about 0.3 kpc in the solar vicinity, the slope of the histogram decreases beyond 0.15 kpc as the shells increasingly overlap with regions outside of the galactic disk. Apart from that, the stellar density decreases away from the central plane of the Milky Way.
- An additional factor is that faint stars at the lower end of the main sequence, which are most abundant in galaxy's stellar population, fall below the limit of 12th magnitude imposed in our sample and, ultimately, below the detection limit of the GAIA telescope.
- Owing to the increasing density of stars towards the galactic center, the number of stars per bin saturates at distances larger than 1 kpc.

Having explained the spatial distribution of the stars in our data set, we can move on to the radial velocity distribution. This time, we explicitly define the edges of the bins, starting from the left edge of the first bin and ending at the right edge of the last bin, in an array named `bins`. The bin width is 2.5 km/s and the left- and rightmost edges limiting the range or radial velocities are -140 and 140 km/s, respectively:

```
17  bin_width = 2.5 # in km/s
18  rv_lim = 140      # upper limit
19  bins = np.arange(-rv_lim, rv_lim+bin_width, bin_width)
20
21  fig = plt.figure(figsize=(6, 4), dpi=300)
22
23  rv_histogram = plt.hist(data[:,3], bins=bins)
24  plt.xlabel('radial velocity / km/s')
25  plt.ylabel('N')
26  plt.savefig('rv_histogram.png')
```

Figure 5.8 shows that the radial velocity distribution appears remarkably like a Gaussian distribution (also called normal distribution) of the form

$$y(x) = y_0 \exp\left(\frac{(x - x_0)^2}{2\sigma^2}\right) \tag{5.7}$$

with an amplitude y_0, mean value x_0, and standard deviation σ. The function $y(x)$ is a so-called probability density function, i.e. it specifies the differential probability of finding the value of the random variable in the infinitesimal interval $[x, x + dx]$. Since the probability that x assumes any value must be unity, normalization of the integral of $y(x)$ implies $y_0 = 1/\sigma\sqrt{2\pi}$. In our case, the random variable x corresponds to the radial velocity. However, the histogram data are not normalized to the total size of the sample. For this reason, we treat y_0 as a free parameter.

In order to verify that the data follow a Gaussian distribution, we compute a fit based on the model given by Eq. (5.7). To that end, we need the bin centers, which can be obtained by shifting the left edges by half of the bin width:

```
27  x = bins[:-1] + bin_width/2
28  y = rv_histogram[0]
```

The bin counts are returned in a row by `plt.hist()` (see line 22). You can easily check that the arrays `x` and `y` have the same size and `x` runs in steps of 2.5 km/s from -138.75 to $+138.75$ km/s:

By applying `curve_fit()` form `scipy.optimize`, we can find which set of parameters y_0, x_0, and σ fits the data best:

```
29  import scipy.optimize as opt
30
31  # definition of fit function
32  def gaussian(x, y0, x0, sigma_sqr):
33      return y0*np.exp(-(x-x0)**2/(2*sigma_sqr))
```

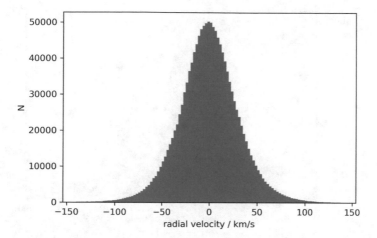

Fig. 5.8 Radial velocity distribution

```
34
35 params, params_covariance = opt.curve_fit(gaussian, x, y)
36 print("Parameters best-fit:", params)
```

Using

```
Parameters best-fit: [ 4.85506870e+04 -8.95860537e-01  7.59083786e+02]
```

we can plot the data together with the fit function:

```
37 y_gauss = gaussian(x, params[0], params[1], params[2])
38
39 fig = plt.figure(figsize=(6, 4), dpi=300)
40
41 plt.hist(data[:,3], bins=bins)
42 plt.plot(x, y_gauss, color='red')
43 plt.xlim(-100,100)
44 plt.xlabel('radial velocity / km/s')
45 plt.ylabel('N')
46 plt.savefig('rv_histo_fit.png')
```

The resulting fit function, which is shown in Fig. 5.9, appears to be in good agreement with the data. Looking closer, you might notice that the core of the distribution is slightly narrower, while the so-called tails toward extreme velocities tend to be broader than the fit function.

Are these deviations just by chance or systematic? Drawing further conclusions requires the application of statistical tests. A common method of calculating the probability that two sets of data, such as observed values and theoretical predictions, follow the same underlying distribution is the Kolmogorov–Smirnov test. This test is implemented in the scipy.stats module:

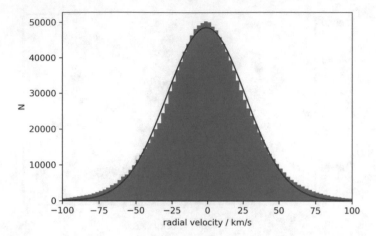

Fig. 5.9 Radial velocity distribution and best-fit Gaussian (red line)

```
47   from scipy.stats import ks_2samp
48
49   ks_2samp(y, y_gauss)
```

And the result is

```
KstestResult(statistic=0.29464285714285715, pvalue=0.00010829928148128113)
```

The famous p-value (printed `pvalue`) is an important measure for statistical significance. In short, assuming that the data follow a given statistical distribution—this is
the so-called *null hypothesis*—the p-value quantifies the probability that a random
sample drawn from the distribution is at least as extreme as the observed data. A very
small p-value indicates that such an outcome is very unlikely if the null hypothesis
were actually true. As a consequence, the null hypothesis can be rejected if the p-
value is very small.[10] In our case, the null hypothesis is that the data are Gaussian.
Since the data appear to deviate from a Gaussian distribution, we test the statistical
significance of our observation. Indeed, we have a p-value as low as 0.01%, which is
well *below* the commonly used threshold of 5%. Therefore, we can conclude that the
distribution of radial velocities in the GAIA data does not strictly follow a Gaussian
distribution, but shows systematic deviations. Gaussian distributions are observed
whenever a certain variable is the result of a superposition of a large number of
random processes. For the solar neighbourhood, the radial velocity of a star relative
to the Sun is determined by many different factors, including the dynamics of the
star-forming region the star originated from or close encounters with other stars. As
we go to larger scales, the contribution of the global galactic rotation profile becomes
more and more important.

---

[10]It is important to be aware that the p-value does not allow us draw conclusions about the likelihood
of any alternative hypothesis.

Galactic dynamics starts to shine through when we take a look at the spatial distribution of radial velocities. We start by dividing the data into two subarrays. Stars with positive radial velocities, which appear red-shifted, are separated from blue-shifted stars with negative radial velocity by means of conditional indexing:

```
50  rv = data[:,3]
51  redshift, blueshift = data[rv > 0], data[rv <= 0]
52
53  print("Redshifted stars:", len(redshift))
54  print("Blueshifted stars:", len(blueshift))
```

```
Redshifted stars: 675676
Blueshifted stars: 710808
```

The number of blueshifted stars is slightly larger than the number of redshifted stars, which suggests a slightly asymmetric distribution.

For a more detailed picture, we produce a scatter plot of red- and blueshifted stars in the plane spanned by the galactic longitude l and latitude b (see Fig. 5.6):

```
55  fig = plt.figure(figsize=(10, 10*60/360+1), dpi=300)
56  ax = fig.add_subplot(111)
57
58  step = 10
59
60  plt.scatter(blueshift[::step,0], blueshift[::step,1],
61              s=1, marker='.', color='blue', alpha=0.1)
62  plt.scatter(redshift[::step,0], redshift[::step,1],
63              s=1, marker='.', color='red', alpha=0.1)
64  plt.xlabel('longitude [deg]')
65  plt.ylabel('lat. [deg]')
66  plt.xlim(0,360)
67  plt.ylim(-30,30)
68
69  # set ticks on axis in 30 degree intervals
70  plt.xticks([30*n for n in range(13)])
71  plt.yticks([-30, 0, 30])
72
73  # ensure that degrees are displayed equally along both axes
74  ax.set_aspect('equal')
75
76  plt.savefig('rv_map.png')
```

We plot only every tenth star. Otherwise the plot would become too crowded with dots. By setting the aspect ratio to 'equal' (the axis object is defined in line 55), we ensure that longitudes and latitudes are shown in proportion. We also set the ticks on the two axes explicitly (see lines 69–70). The plot is shown in Fig. 5.10. You can see that stars tend to be redshifted or blueshifted depending on galactic longitude, while they are randomly scattered along the galactic latitude. The explanation for this variation can be found in the disk-like structure of the Milky Way, with modulations

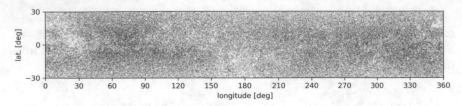

Fig. 5.10 Distribution of blue- and redshifted stars in the *l-b*-plane

caused by its spiral arms. In the following exercise, you are asked to analyse the
dependence on longitude quantitatively.

Exercises

5.5 Use the data set discussed in this section to compute mean values and standard
deviations of the radial velocity for 5° bins of the galactic longitude and plot your
results. If you assume that the galactic rotation curve in the solar neighbourhood is
flat, i.e. the magnitude of the orbital velocity is constant, and the averaged radial
velocities mainly reflect the orbital motions of stars around the center of the galaxy,
which trends do you expect? Keep in mind that the radial velocity is the component
of the velocity difference between a star and the Sun in the direction from the Sun
to the star.

5.6 For distances $d \ll R_0$, where $R_0 \approx 8.5$ kpc is the distance between the Sun and
the center of the Milky way, the radial velocity is approximately given by Oort's
formula

$$\frac{v_r}{d} = A \sin(2l) \,, \tag{5.8}$$

where $A = 14.8$ km s^{-1} kpc^{-1} [4, Sect. 25.3].

(a) Create a scatter plot of v_{rad}/d versus galactic longitude l for those stars in the
     GAIA data set that are closer than $0.05R_0$.
(b) Fit Eq. (5.8) to the data from (a) and determine the best-fit value of A. Add the
     resulting curve to the scatter plot and discuss your results.

5.4   Image Processing

Online access to large collections of images from space telescopes is provided by the
Barbara A. Mikulski Archive for Space Telescopes (MAST) of NASA,[11] including
the legacy archive of the famous Hubble Space Telescope (HST).[12] For example, you
can search for images of the Whirpool Galaxy M51 and download images taken in

---

[11] Archive website: archive.stsci.edu.

[12] See hubblesite.org and www.spacetelescope.org. An example image is shown in Fig. 4.14.

several different wavelength bands from archive.stsci.edu/prepds/m51/datalist.html. These images are available in the FITS file format, which is not only used for data such as spectra, but also for astronomical images. Per se, the CCD cameras used in astronomical telescopes are colour-blind. Every pixel simply counts the number of electrons created by incident photons, regardless of their wavelength. However, the sensitivity of the device is wavelength-dependent.

In order to reconstruct a color image, we need at least three images taken with different wavelength filters. In the following, we select the image data from the blue (435 nm), green/visual (555 nm), and red/Hα (658 nm) filters (in the following also called channels). You need to download the files h_m51_b_s20_drz_sci.fits, h_m51_b_s20_drz_sci.fits, and h_m51_b_s20_drz_sci.fits from the aforementioned URL. We start by reading the FITS header of the red-channel image:

```
1  import numpy as np
2  import matplotlib.pyplot as plt
3  from astropy.io import fits
4
5  m51r_file = "h_m51_h_s20_drz_sci.fits"
6  m51r = fits.open(m51r_file)
7  m51r.info()
```

The info() method shows that the file contains an array of 2150 times 3050 floating point numbers:

```
Filename: h_m51_h_s20_drz_sci.fits
No.    Name       Ver     Type       Cards    Dimensions     Format
  0  PRIMARY        1  PrimaryHDU     1691    (2150, 3050)    float32
```

After copying the data form the primary HDU (see also Sect. 5.1), we can close the file.

```
8  m51r_data = m51r[0].data
9  m51r.close()
```

It is helpful to create a histogram of the image data by flattening m51r_data to a one-dimensional array, i.e. by re-arranging the two-dimensional array of image values in a linear sequence:

```
10  plt.hist(m51r_data.flatten(), log=True, bins=100)
11  plt.xlabel('Signal')
12  plt.ylabel('N')
13  plt.savefig('m51_histogram.png', dpi=300)
```

Since the bin counts N differ by orders of magnitude, we use a logarithmic scale. The histogram shows that by far the largest number of pixels has a signal close to 0, with just a small number of individual readings larger than 5 (see Fig. 5.11). This reason for this becomes clear by viewing the image.

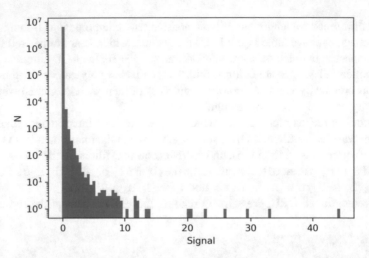

Fig. 5.11 Distribution of values in the M51 red-channel image

To produce a picture, we have to map array values to some color scheme. This is done by the function `imshow()` from `pyplot`. By default, the range between the minimum and maximum is mapped to colors. This would result in an entirely black image as most values are small. For this reason, we set reasonable limits with `clim()`. The upper bound is about ten time the median value, which you can calculate by utilizing NumPy.

```
14  plt.imshow(m51r_data, cmap='gray')
15  plt.clim(0,0.1)
16  plt.colorbar()
17  plt.savefig('m51r.pdf')
```

The image that begins to emerge in Fig. 5.12 is arguably one of the most popular images of a spiral galaxy, sharing many similarities with our own home galaxy, except for the small companion that causes some tidal disruption of the outer spiral arms. However, it still lacks the beautiful light blue of its spiral arms, interlaced by dark dust lanes and reddish star-forming regions.

The image data from the three filters allow us to compose an image using the RGB color system. In this system, colors are represented by a combination of 8-bit values ranging form 0 to 255 for red, green, and blue. First, we complement the image data by loading the green and blue channels:

```
18  m51g_file = "h_m51_v_s20_drz_sci.fits"
19  m51g = fits.open(m51g_file)
20  m51g_data = m51g[0].data
21  m51g.close()
22
23  m51b_file = "h_m51_b_s20_drz_sci.fits"
```

Fig. 5.12 Monochromatic image of M51 in the red (Hβ) channel. The vertical and horizontal axes indicate pixel positions and the color bar on the right shows the mapping of the signal strength to a gray scale

```
24  m51b = fits.open(m51b_file)
25  m51b_data = m51b[0].data
26  m51b.close()
```

The next step is to convert the data into 8-bit values. Since most values are clustered around the mean, we divide the arrays by the mean values in the three channels and multiply by 255. Moreover, we incorporate a factor `alpha` that allows us to shift all channels at once, allowing us to control the brightness of the image.

```
27  alpha = 0.15
28
29  m51rgb = np.zeros([2150, 3050, 3])
30
31  m51rgb[:,:,0] = m51r_data.transpose() / np.mean(m51r_data)
32  m51rgb[:,:,1] = m51g_data.transpose() / np.mean(m51g_data)
33  m51rgb[:,:,2] = m51b_data.transpose() / np.mean(m51b_data)
34
35  m51rgb *= 255*alpha
```

The RGB data are collected in a new, three-dimensional array, where the third dimension spans the red, green, and blue channels. The first two indices specify the position of a pixel, where we have swapped the horizontal and vertical directions by transposing the two-dimensional arrays containing the raw data. As a final step, we cut off at 255 using `np.where()`:

```
36  m51rgb = np.where(m51rgb > 255, 255, m51rgb)
```

Fig. 5.13 RGB image of M51 composed from HST image data from three different filters (archive.stsci.edu/prepds/m51/datalist.html)

Now we have values in the appropriate range. To stack them and turn them into an image, we are going to use the Python Imaging Library (Pillow).[13] The library is important under the name of `PIL`. Among various tools for basic image manipulation and analysis, it provides the function `Image.fromarray()` to create an image object that can be saved in any any common graphics format, such as PNG. Since RGB values must be specified as 8-bit unsigned integers, we apply the method `astype()` to change the array data type from float to `np.uint8` when passing `m51rgb` as argument.

```
37  from PIL import Image
38
39  # convert to 8-bit unsigned integers and turn array into image
40  img = Image.fromarray(m51rgb.astype(np.uint8))
41  img.show()
42  img.save('m51rgb.png')
```

If everything works, you will be greeted by the picture shown in Fig. 5.13. Keep in mind that astronomoical images are to some degree artificial. They do not exactly correspond to the impression the human eye would have if they could see these objects. Even so, they help us to reveal structures such as the prominent star forming regions in M51.

---

[13]For online documentation, see pillow.readthedocs.io/en/stable/handbook/overview.html.

Exercises

5.7 Experiment with the fudge factor `alpha` and the normalization of the image data in the red, green, and blue channels. How does it affect the resulting image of M51? In addition, you can find more image material for merging galaxies under archive.stsci.edu/prepds/merggal.

5.5 Machine Learning

Apart from basic image processing and composition as shown in the previous section, astronomical images are systematically analyzed to infer properties of objects and to classify them. Although humans are very good in recognizing patterns, they are also easily misled by prejudices. To a certain extent, the same could be said about a novel approach to image analysis that, broadly speaking, has become popular under the name of machine learning. Machine learning is based on artificial neural networks (ANNs). In loose analogy to the workings of neurons in a human brain, ANNs are algorithms that can be trained by presenting data to them. The training process enables the network to find features, correlations, structures, etc. when it is confronted with new data. Today, such algorithms are nearly ubiquitous and have become an indispensable tool for the analysis of scientific data, too. The simple use cases demonstrated in this section will hopefully motivate you to seek out other resource to learn more.[14]

A neural network starts with an input layer defining the signals that will be pro cessed by the network. For example, when dealing with image data, the input layer consists of pixel values, such as the 8-bit RGB values introduced in the previous section. In convolutional neural networks (CNNs), which are an advanced type of ANNs, the input layer is followed by a number of convolutional layers in which convolution filters are applied. You can think of them as small windows sliding through an image, combining pixels, and creating feature maps of the inputs (see also Fig. 5.14).

The convolutional layers are so-called hidden layers whose output can be fed to further hidden layers, such as fully connected dense layers performing tasks similar to regression analysis, or to an output layer. A hidden layer consists of artificial neurons that transform an input vector through weights and an activation function into output signals. The weights modulate the signal strength and are adjusted in the process of training the network. The activation function mimics the response of real neurons by introducing non-linearities. A commonly used activation function in multi-layered networks is the rectified linear unit (ReLU), which sets the output signal equal to the weighted sum of the inputs and cuts off at zero. The signals from the last hidden layer are sent to the output layer. The number of neurons in this layer depends on the number of classes that can be identified by the neural network. Machine learning

---

[14]A comprehensive introduction for students is the *Deep Learning* textbook [17]. It is available online: www.deeplearningbook.org.

driven by networks that have arbitrarily complex hidden layers between input and output layer is known as deep learning. It has revolutionized pattern recognition and classification tasks in recent years.

While classical approaches require the programmer to write explicit code for the analysis of specific features, a neural network is trained with large amounts of labelled data and learns to discriminate between different classes of features on its own. This is accomplished by a process called backpropagation: The output the network yields for the training data is compared with the expected output via a loss function, which is a very important choice in the training. Backpropagation calculates the gradient of the loss function with respect to network parameters such as the weights in the hidden layers. The aim is to minimize the loss function by iteratively adjusting the parameters in a number of training epochs. Finally, the network's performance is assessed for independent data sets.

At the end of the day, neural networks boil down to a series of linear and non-linear transformations that are applied to an input vector to turn it into a new vector whose components tell us to which class the input belongs. Thanks to modern APIs such as TensorFlow and Keras,[15] setting up a neural network in Python only takes a few lines of code.[16]

5.5.1 Image Classification

Convolutional neural networks are particularly well suited for the classification of images. An important problem in astronomy is the morphological classification of galaxies. In the following, we will consider only three basic types of galaxies: elliptical, spiral, and irregular.[17] Our goal is to train a CNN using a convolutional layer from the EFIGI catalogue of nearby galaxies so that it will be able to tell us to which of these three classes an arbitrary galaxy belongs [19].

We begin by importing the libraries we will use throughout this section, including the TensorFlow library[18]:

```
1  import numpy as np
2  import matplotlib.pyplot as plt
3  from PIL import Image as image
4  import tensorflow as tf
5  from tensorflow import keras
```

---

[15] API stands for application programming interface. It is something like a construction kit that makes the programming of applications easier for you.

[16] See [18] for a practical, Python-based guide to machine learning.

[17] A more detailed morphological classification scheme for galaxies is the so-called Hubble sequence.

[18] A good start point is www.tensorflow.org/lite/guide. TensorFlow is not a standard package in Python distributions. In Anaconda, for example, you need to run `conda install tensorflow` before you can use the package.

All information we need to work with the sample of EFIGI galaxy images is stored in the `efigi.dat` file. The first column contains the filename of every single image and the second column the morphological class it belongs to. Here, we use Python lists and the built-in function **open()**:

```
6   data = open("data_files/galaxies/efigi.dat","r")
7
8   names = []
9   types = []
10
11  for line in data:
12      fields = line.split(" ")
13      names.append( fields[0] )
14      types.append( fields[1] )
15
16  nData = len(names)
17  imgSize = 64
```

The lists `names` and `types` are created as empty lists, i.e. initially they have no elements. With each iteration of the subsequent loop, a line is read from the file and the extracted name and type are appended as new elements to the lists.

The original images are 255×255 pixels in size, but we will scale them down to `imgSize` to reduce the computational cost. The galaxies are stored in a multi-dimensional array whose first dimension is `nData` (number of images), the second and third dimensions are both `imgSize` (corresponding to 64×64 pixel values), and the fourth dimension corresponds to the three color channels of an RGB image (see Sect. 5.4):

```
18  galaxies = np.zeros((nData, imgSize, imgSize, 3))
19  labels = np.zeros(nData, dtype='int')
20
21  for i in range(nData):
22      # load image
23      img = image.open("data_files/galaxies/png/" +
24                       str(names[i]) + ".png")
25
26      # resize to imgSize
27      imgResized = img.resize(size-(imgSize,imgSize))
28
29      galaxies[i,:,:,:] = np.array(imgResized)/255
30      labels[i] = types[i]
```

In line 23, the image files are opened using the module `Image` form PIL. Each image is rescaled to the size given by `imgSize`, converted to a numpy array, and its pixel values are normalized to unity.

In addition to a training set, we need two smaller convolutional layers for validating and testing the network performance. Validation proceeds parallel to the training process and helps to cross-check whether the network tends to memorize special features of the training data, while failing to identify general features in an independent data set. The test convolutional layer is used for a final performance test. To avoid any biases, we need to make sure that the three convolutional layers are random samples of the available data. For this reason, we first split the data into a training set, which usually encompasses about 70% of all data, and then subdivide the remainder into a validation and a test set:

```
31  import random
32
33  # generate random sample of unique indices
34  size = labels.size
35  sample = random.sample([n for n in range(size)], int(0.3*size))
36
37  # split into training and other set otherLabels = labels[sample]
38  otherGalaxies = galaxies[sample,:,:,:] trainLabels =
39  np.delete(labels, sample) trainGalaxies = np.delete(galaxies,
40  sample, axis=0)
41
42  print(otherLabels.size, trainLabels.size)
43  print(otherGalaxies.shape, trainGalaxies.shape)
```

By means of `random.sample()`, we randomly select 30% of all indices of the `labels` array without duplicates. The resulting list can be used for indexing.[19] For example, `otherLabels` defined in line 38 is an array containing only those elements of `labels` with indices included in the list `sample`. This is similar to extracting elements by slicing, except for specifying a list of indices instead of an index range. The complement is obtained by deleting all elements indexed by `sample` (see line 40). We are left with two arrays of sizes

```
322 754
```

The same operations are applied to axis 0 (i.e. the first dimension running through all images) of the `galaxies` array, producing two arrays with the following shapes:

```
(322, 64, 64, 3) (754, 64, 64, 3)
```

These arrays are in turn split into the validation and test samples, each containing one half of the elements (or 15% with respect to the original size):

```
44  size = otherLabels.size
45  subsample = random.sample([n for n in range(size)],
46                            int(size/2))
47
48  # split into validation and test sets
49  valdLabels = otherLabels[subsample]
50  valdGalaxies = otherGalaxies[subsample,:,:,:]
51  testLabels = np.delete(otherLabels, subsample)
52  testGalaxies = np.delete(otherGalaxies, subsample, axis=0)
```

---

[19] Alternatively, an array may be indexed with an integer array.

You may want to check that the samples have similar distributions of ellipticals, spirals, and irregulars by plotting histograms of the three label arrays.

The next step is to create the network. The architecture of a neural network that fits a given problem is—to a certain degree—the result of a trial and error process. Is a single hidden layer sufficient or do we need more than one? How many convolutional layers of what size do we need? For some problems, working networks have already been proposed and can be used as a starting point.[20] Let us try the relatively simple network illustrated in Fig. 5.14:

```
53  galNet = keras.Sequential([
54      keras.layers.Conv2D(96, (8,8), activation='relu',
55                          input_shape=(imgSize,imgSize,3)),
56      keras.layers.MaxPooling2D(pool_size=(4,4)),
57      keras.layers.Flatten(),
58      keras.layers.Dense(30, activation='relu'),
59      keras.layers.Dense(3, activation='softmax')
60  ])
```

The network is created by the `Sequential` function defined in the `keras` module.[21] The individual layers are specified in the argument list. The first layer creates 96 feature maps by applying convolution filters to image arrays with dimensions defined by `input_shape`. The kernel of each filter is given by a matrix with 8×8 elements. To reduce the amount of data generated by this process, the convolutional layer is followed by a pooling layer that reduces the size of the data by downsampling the feature maps. This completes the feature extraction. The resulting data must be flattened into a one-dimensional vector before they can be processed by a fully-connected dense layer consisting of 30 neurons. The term fully-connected refers to the property of all inputs being connected to every neuron. The signals from the neurons are distributed to three output nodes that represent the three classes we are dealing with. By using the so-called softmax activation, we obtain probabilistic results. The network structure is summarized by `galNet.summary()`:

Layer (type)	Output Shape	Param #
conv2d_6 (Conv2D)	(None, 57, 57, 96)	18528
max_pooling2d_6 (MaxPooling2	(None, 14, 14, 96)	0
flatten_6 (Flatten)	(None, 18816)	0
dense_12 (Dense)	(None, 30)	564510
dense_13 (Dense)	(None, 3)	93

Total params: 583,131
Trainable params: 583,131
Non-trainable params: 0

---

[20]For example, see arxiv.org/abs/1709.02245 for a simple application of CNNs to galaxy classification.

[21]See www.tutorialspoint.com/tensorflow/tensorflow_keras.htm.

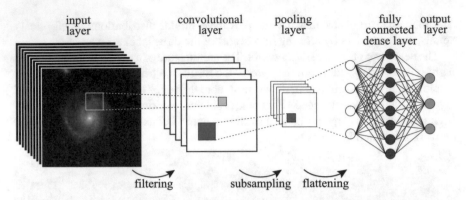

Fig. 5.14 Schematic view of a convolutional neural network (Galaxy image from the EFIGI catalogue [19])

At the bottom, you can see that a total number of 583,131 parameters have to be optimized!

Before we can train our network to learn which input signals are typical for which output class, we need to compile it:

```
61  galNet.compile(optimizer='adam',
62                 loss='sparse_categorical_crossentropy',
63                 metrics=['accuracy'])
```

The `optimizer` keyword specifies the numerical optimization algorithm that will be used to adjust the weights in the process of training the network. The sparse categorical crossentropy loss function we are going to use is a standard choice when dealing with categorical data, where every item belongs to one category. The third argument, with the keyword `metrics`, indicates that we will use the fraction of correctly classified training images to score the performance of the network.

Now we are all set to train the network by applying the `fit()` method to the training sample for a given number of epochs. The validation sample can be passed as optional argument with the keyword `validation_data`:

```
64  results = galNet.fit(trainGalaxies, trainLabels, epochs = 40,
65                  validation_data=(valdGalaxies,
66                                   valdLabels))
```

On a present-day PC, the training process should not take more than a few minutes. Once we are through, we can take a look at the evolution of the network's performance as a function of the number of training epochs (i.e. the number of iterations). The losses of the two samples are returned as items in the `history` dictionary:

```
67  plt.figure(figsize=(6,4), dpi=100)
68
69  plt.plot(results.history['loss'], color='green',
```

```
70            label='training')
71  plt.plot(results.history['val_loss'], color='red',
72            label='validation')
73  plt.xlabel("Epochs")
74  plt.ylim(0,1)
75  plt.ylabel("Loss")
76  plt.legend()
77  plt.savefig("galnet_loss.pdf")
```

As expected, the loss curve for the training data set descends with the number of epochs (see Fig. 5.15). However, the loss for the validation set levels off and even increases after 20 or so epochs. This is a symptom of overfitting: Under certain conditions, the network does not recognize general features that characterize a class, but simply memorizes every specific element in the training data set. This is also reflected by the accuracies at the end of the training:

```
78  print(f"{results.history['accuracy'][-1]:.4f} "
79        f"{results.history['val_accuracy'][-1]:.4f}")
```

```
    0.9947 0.8634
```

Hence, labels in the training sample are much better reproduced than in the validation sample.

Overfitting happens when a network has too many parameters. In other words, it is too complex in relation to the size and diversity of the data. There are several ways how to avoid overfitting. The most obvious is to collect more data for the training of the network. If we need to content ourselves with the data we have, we can try to reduce the number of parameters, e.g. by using a smaller number of feature maps. Another option is to reduce the number of fully connected neurons. However, rather than just reducing the number of neurons it is often preferable to insert a so-called dropout layer that randomly deactivates neurons:

```
80  galNet = keras.Sequential([
81      keras.layers.Conv2D(32, (8,8), activation='relu',
82                          input_shape=(imgSize,imgSize,3)),
83      keras.layers.MaxPooling2D(pool_size=(4,4)),
84      keras.layers.Flatten(),
85      keras.layers.Dropout(0.3),
86      keras.layers.Dense(24, activation='relu'),
87      keras.layers.Dense(3, activation='softmax')
88  ])
```

The summary shows that the number of parameter is reduced considerably:

Layer (type)	Output Shape	Param #
conv2d_1 (Conv2D)	(None, 57, 57, 32)	6176

```
max_pooling2d_1 (MaxPooling2  (None, 14, 14, 32)              0
```
```
flatten_1 (Flatten)           (None, 6272)                    0
```
```
dropout (Dropout)             (None, 6272)                    0
```
```
dense_2 (Dense)               (None, 24)                 150552
```
```
dense_3 (Dense)               (None, 3)                      75
=================================================================
Total params: 156,803
Trainable params: 156,803
Non-trainable params: 0
```

After compiling the network and running `galNet.fit()` again, we can investigate the loss curves. The results are shown in Fig. 5.15. The divergence between the training and validation losses is reduced and the accuracies after the final training epoch are not as far apart as before:

```
0.9735 0.8882
```

You can try to further improve the network (see Exercise 5.8).

We finish the training at this point and evaluate the accuracy for the test data set:

```
89  loss, acc = galNet.evaluate(testGalaxies, testLabels)
```

The result for our sample (value of `acc`) shows that the network is capable of classifying 86% of all test images correctly. This is a really good score for a rather simple CNN and a data set of moderate size. Although the accuracy has not improved significantly compared to the validation of the larger network, the smaller network has the advantage of being computationally less expensive. This means it is adequate for the given data set.

Now that our network is trained, we can use the `predict()` method to query individual images or batches of images. As an example, we load an HST image of NGC 1232 and convert it into an image array[22]:

```
90  img = image.open("data_files/galaxies/NGC_1232.jpg")
91
92  imgResized = img.resize(size=(imgSize,imgSize))
93
94  imgArr = np.array(imgResized)/255
```

Since `predict()` expects an array whose shape matches the training data set (except for the size of the data set), we need to expand `imgArr` to a four-dimensional array, i.e. an array with four indices, by inserting a new axis:

---

[22] You can download the image from commons.wikimedia.org/wiki/File:NGC_1232.jpg.

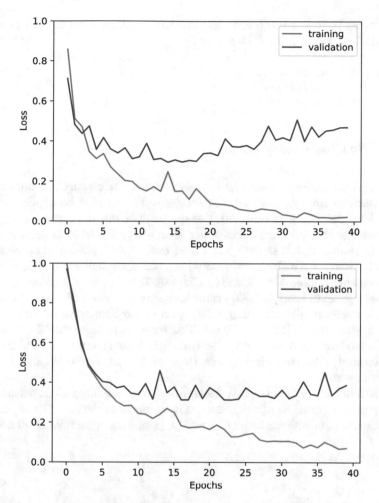

Fig. 5.15 Loss curves of simple galaxy classification networks with 96 feature maps (top) and 32 feature maps as well as a dropout layer (bottom) for training and validation data

```
95  imgArrExp = np.expand_dims(imgArr, axis=0)
96  print(imgArrExp.shape)
```

```
(1, 64, 64, 3)
```

Since we have only one image, the expanded array has, of course, only one element along the new axis. Let us see what we get:

```
97  pred = galNet.predict(imgArrExp)
98
99  label = ["elliptical", "spiral", "irregular"]
00  for i,p in enumerate(pred.flatten()):
01      print(f"{label[i]:10s} {p:.4e}")
```

The network is almost 100% confident that NGC 1232 belongs to class of spirals (second label), which is indeed the case:

```
elliptical 3.0359e-10
spiral     9.9963e-01
irregular  3.6675e-04
```

5.5.2 Spectral Classification

In the previous section, we have set up a network that can classify two-dimensional inputs such as images. Naturally, neural networks can also be applied to one-dimensional data, such as time series or spectra. We will show you how to build a network that can derive the effective temperature of a star from its spectrum. This would be an easy task if stellar spectra were exact Planck black-body spectra (see Sect. 3.1). However, a multitude of absorption processes in stellar atmospheres makes real spectra much more complicated [3, Chap. 9]. To train our network, we are going to use a large set of nearly 80,000 synthetic spectra with overlaid noise,[23] each one stored in a separate file containing 8,500 spectral flux density values in the wavelength range between 585 and 670 nm. This amounts to more than 2 GB of data, which is too large for download via the URL from which source code and the other data files used in this book are available. However, the data can be obtained from the authors on request.

In the following, we assume that the data files for the training are located under the path `specnet/training` relative to the work directory.[24] The os library enables us to scan the path with the help of `listdir()` and add the names of all files to a list:

```
1   import numpy as np
2   import matplotlib.pyplot as plt
3   import tensorflow as tf
4   from os import listdir
5   from os.path import isfile, join
6
7   path = "specnet/training"
8   specnames = [f for f in listdir(path) if isfile(join(path, f))]
9
10  n_spectra = len(specnames)
11  print("Total number of training spectra:", n_spectra)
```

In line 8, the file list is generated with an inline loop combined with a conditional statement that checks whether an entry found by `listdir()` is a file or not. The number of files corresponds to the number of different spectra:

---

[23] Synthetic spectra are computed using stellar atmosphere codes. The spectra in our training set were produced from a small sample of computed spectra, which were overlaid with a large number of different realizations of random noise to mimic instrumental noise in observed spectra. The original synthetic spectra were produced with the Spectroscopy Made Easy (SME) package [20].

[24] It might be necessary to adjust the path depending on your operating system and the location of the specnet directory on your computer.

```
    Total number of training spectra: 79200
```

The file names (e.g. 5800_1_65999_177.97.npz) contain the effective temperatures of the stars, defining the data labels for the training of the network. For this reason, we extract the first four digits from each file name and save them in the list temp. As each temperature occurs many times, we create an ordered list of unique temperature classes (in the astrophysical sense) by applying a combination of the functions **set**() (creates an unordered set of items), **list**() (converts back to a list), and **sorted**() (arranges the list elements in ascending order):

```
12  temp = np.zeros(n_spectra, dtype='int')
13
14  for i,spec in enumerate(specnames):
15      temp[i] = int( spec[0:4] )
16
17  temp_class = sorted(list(set(temp)))
18  n_labels = len(temp_class)
19
20  print("Total number of temperature classes:", len(temp_class))
21  print("List of temperatures:", temp_class)
```

Here we go:

```
    Total number of temperature classes: 11
    List of temperatures: [4000, 4200, 4400, 4600, 4800, 5000, 5200, 5400, 5600, 5800, 6000]
```

Thus, our data set encompasses spectra for effective temperatures in the range from 4000 to 6000 K, corresponding to stars in the spectral classes K and G [4, 21].

Let us take a closer look at the individual spectra. The suffix .npz implies that we are dealing with zipped binary NumPy arrays. They can be loaded with np.load(). In the following code example, an arbitrarily chosen spectrum is loaded into an array:

```
22  spectrum_file = join(path, "5800_1_65999_177.97.npz")
23
24  spec_arr = np.load(spectrum_file)
25  print(spec_arr.files)
```

The output from the last lines shows that the file contains just a single array, which is referenced by the keyword "arr_0":

```
    ['arr_0']
```

For simple use, we slice spec_arr into two new arrays, namely wave representing the wavelengths and flux holding the spectral flux densities

```
26  wave = spec_arr["arr_0"][:,0]
27  flux = spec_arr["arr_0"][:,1]
28
29  print("Wavelength range:", np.min(wave), np.max(wave))
30
31  spec_size = len(flux)
32
33  print("Number of values per spectrum:", spec_size)
```

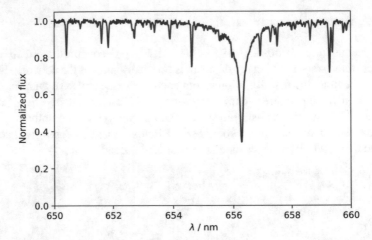

Fig. 5.16 Section of a synthetic spectrum for $T_{eff} = 6000$ K. The small fluctuations in the continuum between absorption lines stem from overlaid noise

```
Wavelength range:  5850.0 6700.0
Number of values per spectrum: 8500
```

Figure 5.16 shows the spectrum in the wavelength range from 650 to 660 nm (wavelengths printed above are in angstrom). Are you able to identify the prominent absorption line at about 656 nm?

All in all we are confronted with a total of 672,800,000 spectral data points. This is a huge number. For computational purposes, we subdivide the wavelength range into 20 channels of equal size:

```
34  n_channels = 20
35  channel_length = int(spec_size/n_channels)
36
37  print("Values per channel:", channel_length)
```

```
Values per channel: 425
```

Since TensorFlow expects all of our input data to be contained in a single array, we need to iterate through the file list and successively store the spectral data in an array whose first dimension equals the number of spectra. The sequential data for each spectrum are reshaped so that the second index of the data array runs through the channel specified by the third index. Moreover, we create an array in which the labels will be stored. Since the class labels must be specified as integers, the temperature class of each spectrum is mapped to an index via the index() method (see line 42). A word of caution: You need a computer with sufficient main memory and speed to proceed from here on.

```
38  labels = np.zeros(n_spectra, dtype='int')
39  spectra = np.zeros((n_spectra, channel_length, n_channels),
40                     dtype='float64')
41
```

```
42   for i in range(n_spectra):
43       labels[i] = temp_class.index(temp[i])
44
45       spectrum_file = join(path, specnames[i])
46       spec_arr = np.load(spectrum_file)
47
48       flux = spec_arr["arr_0"][:,1]
49       flux_2d = np.reshape(flux, (-1,n_channels))
50
51       spectra[i,:,:] = flux_2d
52
53   print(spectra.shape)
```

After loading all data, we have an impressive array with the following dimensions:

```
(79200, 425, 20)
```

Similar to the CNN for image classification in the previous section, we can set up a network to classify the spectra, with the important difference of using one-dimensional convolutional layers:

```
54   specNet = tf.keras.models.Sequential([
55       tf.keras.layers.Conv1D(24, 4, activation='relu',
56           input_shape=(channel_length, n_channels)),
57       tf.keras.layers.Conv1D(120, 10, activation='relu'),
58       tf.keras.layers.Flatten(),
59       tf.keras.layers.Dense(n_labels, activation='softmax'),
60   ])
61
62   print(specNet.summary())
```

The input_shape parameter of the first layer is given by the channel length (i.e. number of wavelength bins per channel) and the number of channels. Nevertheless, the first convolutional layer is one-dimensional, as indicated by Conv1D(), and the kernel has only a single dimension. In addition, we use a second convolutional layer with a larger number of filters, but there is no dense layer apart from the output layer. All the magic happens in the convolutional layers. Here is a summary of the chosen CNN:

```
Model: "sequential"
```

Layer (type)	Output Shape	Param #
conv1d (Conv1D)	(None, 422, 24)	1944
conv1d_1 (Conv1D)	(None, 413, 120)	28920
flatten (Flatten)	(None, 49560)	0
dense (Dense)	(None, 11)	545171

```
Total params: 576,035
Trainable params: 576,035
Non-trainable params: 0
```

Unfortunately, there is no general recipe how to build a network. It requires intuition and repeated cycles of training, performance evaluation, and parameter tuning.

The network defined above is compiled just like in the previous section:

```
63  specNet.compile(optimizer='adam',
64                  loss='sparse_categorical_crossentropy',
65                  metrics=['accuracy'])
```

Owing the very large data size, we go only through four training cycles[25]:

```
66  specNet.fit(spectra, labels, epochs=4)
```

```
Epoch 1/4
2475/2475 [==========================] - 29s 12ms/step - loss: 0.3448 - accuracy: 0.8878
Epoch 2/4
2475/2475 [==========================] - 29s 12ms/step - loss: 0.0927 - accuracy: 0.9755
Epoch 3/4
2475/2475 [==========================] - 29s 12ms/step - loss: 0.0516 - accuracy: 0.9972
Epoch 4/4
2475/2475 [==========================] - 29s 12ms/step - loss: 0.0013 - accuracy: 1.0000
```

As you can see from the output, four epochs are sufficient to reach a very high accuracy. However, we left out validation (see Exercise 5.9).

The test data set under the path specnet/training can be loaded analogous to the training data set (the code is omitted here). How does our network perform?

```
67  loss, acc = specNet.evaluate(spectra_test, labels_test)
68
69  print(f"Accuracy: {acc:.4f}")
```

It turns out that the result is perfect:

```
Accuracy: 1.0000
```

After learning the main features of the 11 temperature classes, the network is able to correctly classify 100% of our test data. We can take a closer look on the network's classification confidence by inspecting the prediction for a single spectrum:

```
70  i_test = 4000
71  print("Name of the spectrum:", specnames_test[i_test], "\n")
72
73  spec = spectra_test[i_test]
74
75  spec_exp = np.expand_dims(spec,0)
76
77  guess = specNet.predict(spec_exp)
78
79  for i in range(n_labels):
80      print("{:4d} K  {:6.2f} %".
81            format(temp_class[i], 100*guess[0,i]))
```

---

[25]Timings are from a run using an Intel i7 CPU. We also used a Nvidia Titan V graphics card (GPU) to accelerate the training (see Exercise 5.9 for GPU offloading). This allowed us to complete the training in less than 40 s.

```
Name of the spectrum: 4200_1_1101_698.97.npz
```

```
4000 K      0.00 %
4200 K    100.00 %
4400 K      0.00 %
4600 K      0.00 %
4800 K      0.00 %
5000 K      0.00 %
5200 K      0.00 %
5400 K      0.00 %
5600 K      0.00 %
5800 K      0.00 %
6000 K      0.00 %
```

The network finds by far the highest probability for the correct answer: 4200 K.

It is good practice to save the model, i.e. the trained network with all parameters, so that it can be used anytime later without repeating the training all over again:

```
82  SpecNet.save('data_files/specnet_model.tf', save_format='tf')
```

This file is part of this chapter's online material. You can restore the network with

```
83  specNet = tf.keras.models.load_model('data_files/specnet_model.tf')
```

Since we used synthetic spectra for training and testing, the crucial question is whether the network will work with real data from observations. As a demonstration, let us apply our network to the solar spectrum:

```
84  spectrum_file = "data_files/sun_spec.npz"
85
86  spec_arr = np.load(spectrum_file)
87  wave = spec_arr["arr_0"][:,0]
88  flux = spec_arr["arr_0"][:,1]
89
90  flux_2d = np.reshape(flux, (-1,n_channels))
```

If you like you can plot the spectrum. Now let us see what specNet's guess is:

```
91  guess = specNet.predict(np.expand_dims(flux_2d, axis=0))
92
93  for i in range(n_labels):
94      print("{:4d} K  {:6.2f} %".
95              format(temp_class[i], 100*guess[0,i]))
96
97  print("\nEffective temperature estimate: {:.0f} K".
98          format(np.average(temp_class, weights=guess.flatten())))
```

Table 5.1 Spectral classes and effective temperatures of G- and K-type main sequence stars

Class	T_{eff} [K]
G0	5980
G2	5800
G5	5620
G9	5370
K0	5230
K1	5080
K3	4810
K4	4640
K5	4350
K7	4150

```
4000 K      0.00 %
4200 K      0.00 %
4400 K      0.00 %
4600 K      0.00 %
4800 K      0.00 %
5000 K      0.00 %
5200 K      0.00 %
5400 K      0.00 %
5600 K      1.15 %
5800 K     98.81 %
6000 K      0.03 %


Effective temperature estimate: 5798 K
```

The weighted average of 5798 K closely matches the Sun's effective temperature of 5780 K. Our network has indeed learned to estimate the effective temperature of a star from its spectrum.

From the use cases discussed in this section, image-based galaxy classification and spectral classification of stars, you can guess how powerful CNNs are. But we have only scratched at the surface of machine learning. By now, there is a plethora of applications in astrophysics. It is up to you to explore these topics further.

Exercises

5.8 Are you able to improve `galNet`? Vary the network parameters of the convolution and dropout layers. If your computer permits, you can also attempt to use a larger image size for the training or experiment with an additional convolutional layer. Once you are finished with validation and testing, search for galaxy images on the web and feed them into your network.

5.9 Instead of temperatures, use the spectral classes according to the Morgan–Keenan system to label your data (see Table 5.1). For example, the spectral class of the Sun is G2. Label each synthetic spectrum with the class that is closest to its effective temperature (the lowest temperatures fall in the same class). Train the network to identify these classes and investigate the sensitivity on the filter parameters of the second convolutional layer. With the keyword `validation_split` you can use a fraction of the training data set as validation data to compare losses in the training epochs.

If you have a powerful graphics card, you can try to offload the training by using TensorFlow's `device()` function:

```
with tf.device('/gpu:0'):
    specNet.fit(spectra, labels, epochs=4)
```

The `with` statement creates a so-called context manager for the execution of the training on the GPU of your computer (see also Appendix B.3 and the tutorial www.tensorflow.org/guide/gpu).

Appendix A
Object-Oriented Programming in a Nutshell

Classical programming languages, such as Fortran or C, are procedural languages: Especially in low-level languages like C, every statement corresponds to one or several lines of assembly, which will in turn be translated into a number of native CPU instructions. Variables and functions (subroutines) acting on variables are strictly distinct concepts: The former are basically locations in memory, the latter collections of instructions that can be called repeatedly.

The late 20th century has seen the rise of a new programming paradigm: Object-oriented programming (OOP) languages aim at bringing together data and functions by introducing the concept of classes. A class is a blueprint for objects (also known as instances) containing specific data and functions to manipulate the data in a controlled way. Thus, data are *encapsulated* in objects, making code easier to read and to understand. Generally, object-oriented languages are well suited for developing complex applications that do not crave for every single bit of performance and do not require low-level access to hardware. Python is not a purely object-oriented programming language (you can write embarrassingly procedural code in Python), but it supports the basic concepts of OOP. Apart from encapsulation, these are inheritance, abstraction, and polymorphism. While we briefly touch upon inheritance in Sect. 2.1.3, abstraction and polymorphism are not covered by this book.

Let us consider an example: In Sect. 4.5, we analyze data from an N-body simulation of a stellar cluster. The basic entities in a gravitational N-body code are the bodies, i.e. point particles of a certain mass that reside at some position and move with some velocity. Mass, position, and velocity are attributes of the particles. Given these data, we can evaluate the instantaneous gravitational forces between the bodies and change their position and velocities using a numerical integration scheme for differential equations. In a more advanced code, bodies may have additional attributes, for example, a finite radius for Roche limit calculations. We will not go into the difficulties of implementing full N-body code here, but we will consider the simplest case, namely the two body-problem, to explain some of the basic ideas.

© Springer Nature Switzerland AG 2021
W. Schmidt and M. Völschow, *Numerical Python in Astronomy and Astrophysics*,
Undergraduate Lecture Notes in Physics,
https://doi.org/10.1007/978-3-030-70347-9

An essential part of a Python class is the definition of the __init__ method, which is also called constructor. It defines the attributes of an object belonging to the class. Below the complete definition of the class Body is listed. It is part of the module nbody, which is zipped together with other material from the appendices of this book. The class definition begins with the Python keyword **class** followed by the name of the class. The so-called member functions of the class are indented:

```python
1   # excerpt from nbody.py
2   import numpy as np
3   from scipy.constants import G
4
5   class Body:
6
7       location="Universe"
8
9       def __init__(self, m, name=None):
10          self.m = m
11          self.name = name
12
13          # protected attributes
14          self._x = np.zeros(3)
15          self._v = np.zeros(3)
16
17      def print_mass(self):
18          if self.name == None:
19              print(f"Mass = {self.m:.2e} kg")
20          else:
21              print("Mass of", self.name, f"= {self.m:.2e} kg")
22
23      def set_state(self, x0, v0):
24          # ensure x0 and v0 are arrays
25          x0 = np.array(x0); v0 = np.array(v0)
26
27          # accept only if there are three elements
28          try:
29              if x0.size == 3 and v0.size == 3:
30                  self._x = x0
31                  self._v = v0
32              else:
33                  raise ValueError
34          except ValueError:
35              print("Invalid argument:",
36                    "must be array-like with three elements")
37
38      def pos(self):
39          return self._x
40
41      def vel(self):
42          return self._v
43
44      # compute distance between this body and another
```

```
45      def distance(self, body):
46          try:
47              if isinstance(body, Body):
48                  return ((self._x[0] - body._x[0])**2 +
49                          (self._x[1] - body._x[1])**2 +
50                          (self._x[2] - body._x[2])**2)**(1/2)
51              else:
52                  raise TypeError
53          except TypeError:
54              print("Invalid argument:",
55                      "must be instance of Body")
56
57      # compute gravitational force exerted by another body
58      def gravity(self, body):
59          delta =  body._x - self._x # distance vector
60          return G * self.m * body.m * \
61              delta / np.sum(delta*delta)**(3/2)
62
63      @classmethod
64      def two_body_step(cls, body1, body2, dt):
65          """
66          symplectic Euler step for the two-body problem
67
68          args: body1, body2 - the two bodies
69                dt - time step
70          """
71          force = cls.gravity(body1, body2)
72
73          body1._v += force * dt / body1.m
74          body2._v -= force * dt / body2.m
75
76          body1._x += body1._v * dt
77          body2._x += body2._v * dt
```

The constructor is defined in lines 9–15. To create a new instance of the class, the user calls the constructor and supplies all the information that is required to initialize the attributes. But how does this work?

In Sect. 4.3, we defined initial conditions for the binary system Sirius A and B. So, let us first create an object for the star Sirius A:

```
1  %load_ext autoreload
2  %autoreload 1
3  %aimport nbody
4
5  from astropy.constants import M_sun
6
7  body1 = nbody.Body(2.06*M_sun.value, "Sirus A")
```

The first step is, of course, to import the module nbody, which you would normally do by executing **import** nbody. In interactive Python, however, class definitions contained in modules will not be updated once you have imported the module, unless it is explicitly reloaded. This can be a quite a nuisance when you are still in the process of developing a class. Fortunately, IPython and Jupyter offer with the autoreload extension a convenient gadget. After invoking the magic command %autoreload 1, all modules imported with %aimport will be reloaded whenever subsequent code is executed. In the example above, nbody is automatically reloaded and you will find that any changes to nbody.py made in a source code editor will immediately come into effect in your interactive session. After loading the required modules, an object called body1 is created in line 7. As you can see, the constructor is called by the class name and its first argument, self, is omitted. In Python classes, the self variable represents the current object and only has a meaning inside a class. In the constructor it refers to the object to be created. For this reason, it cannot be specified as an actual argument. In other words, __init__(self) is equivalent to Body() outside of the the class definition. The other arguments set the mass and the name of the star as attributes. The constructor also initializes the position and velocity with null vectors (defined as NumPy arrays).

Object attributes can be easily accessed via the dot operator. For example,

```
8  print("Mass of", body1.name, f"= {body1.m:.2e} kg")
```

produces the formatted output

```
    Mass of Sirus A = 4.10e+30 kg
```

However, only *public* attributes should be accessed this way. There are also *protected* and *private* attributes, but Python tends to be less restrictive than other object-oriented languages adhering to the principle of data encapsulation. By default, attributes are public and you are allowed to modify them directly. You may want to give it a try. While it makes sense to change basic properties such as the name of a body without further ado, position and velocity are more delicate pieces of data. They change according to physical laws. For this reason, the attributes _x and _v are prefixed by an underscore, indicating that the user is not supposed to access them outside of the class. You can change them directly, but the preferred way of access is via methods.

Such attributes are called protected.[1] If an attribute is prefixed with two underscores, it is private and access will be more restricted.

Printing the mass of a body is incorporated as a method in the class `Body` (see listing above). As a result, it can be applied to any object:

```
9   body2 = nbody.Body(1.02*M_sun.value, "Sirus B")
10
11  body2.print_mass()
```

In this case, we get the output

```
Mass of Sirus B = 2.03e+30 kg
```

From the definition of `print_mass()` you can see that the output is adjusted to the cases of a name being defined or not. The `name` attribute is an optional argument of the constructor. By default it is set to `None`. You can define a name anytime later if you so choose. Maybe you remember various examples where whole objects are printed (see, for example, Sect. 1.4). This is the purpose of the `__str__(self)` method. It returns a formatted string for printing some or all attributes of an object in a well readable from (e.g. **print**(body2)) We leave it to you to add such a method to the `Body` class.

The next step is to define initial data. This can be done along the same lines as in Sect. 4.3, except for using the `set_state()` method to set the initial position and velocity of each star:

```
12  from math import pi
13  from scipy.constants import au,G
14
15  M1 = body1.m
16  M2 = body2.m
17
18  # orbital parameters
19  a = 2.64*7.4957*au
20  e = 0.5914
21  T = pi * (G*(M1 + M2))**(-1/2) * a**(3/2)
22
23  # periastron
24  d = a*(1 - e)
25  v = (G*(M1 + M2)*(2/d - 1/a))**(1/2) # vis-viva eq.
26
27  body1.set_state([d*M2/(M1 + M2), 0], [0, -v*M2/(M1 + M2)])
```

The last line, however, will throw an error:

```
Invalid argument: must be array-like with three elments
```

The reason is that `set_state()` checks whether the arguments it receives are arrays with three elements. If not, a `ValueError` is raised. The problem in

---

[1]This is a naming convention. Protected attributes can make a difference if subclasses are introduced.

the example above is that two-dimensional position and velocity vectors are passed as arguments. Although this makes sense for the two-body problem, the general case with three dimensions is assumed in the class Body. Consequently, the correct initialization reads

```
28  body1.set_state([d*M2/(M1 + M2), 0, 0],
29                  [0, -v*M2/(M1 + M2), 0])
30  body2.set_state([-d*M1/(M1 + M2), 0, 0],
31                  [0, v*M1/(M1 + M2), 0])
```

You can get, for example, the position of Sirius A via body1.pos().

To compute the initial distance of the two stars, we can use another method:

```
32  print("{:.2f} AU, {:.2f} AU".
33        format(d/au, body1.distance(body2)/au))
```

This confirms that the initial distance is the periastron distance defined in line 24:

```
    8.09 AU, 8.09 AU
```

Since distance() is a method, the first body is self, while the other body is passed as an argument (see the definition of the method in the listing above). To make the method foolproof, it is checked whether the actual argument is an instance of the class. Actually, there is a way to call an *instance* method with the self argument being filled in by the argument list:

```
34  print(nbody.Body.distance(body1, body2)/au)
```

You would not normally want to do that, but there are exceptions from the rule (see below).

Finally, let us put all pieces together and simulate the orbital motions of Sirius A and B:

```
35  import numpy as np
36
37  n_rev = 3        # number of revolutions
38  n = n_rev*500    # number of time steps
39  dt = n_rev*T/n   # time step
40  t = np.arange(0, (n+1)*dt, dt)
41
42  orbit1 = np.zeros([n+1,3])
43  orbit2 = np.zeros([n+1,3])
44
45  # integrate two-body problem
46  for i in range(n+1):
47      orbit1[i] = body1.pos()
48      orbit2[i] = body2.pos()
49
50      nbody.Body.two_body_step(body1, body2, dt)
```

In the `for` loop, we make use of the *class* method `two_body_step()` to update positions and velocities using the symplectic Euler scheme. In contrast to an instance method, it does not act on a particular object. In the above listing of the class, you can see that the definition of `two_body_step()` is marked by the `@classmethod` decorator. The argument named `cls` is analogous to `self`, except that it refers to the class name instead of the current object. The two objects for which we want to compute the Euler step are passed just like arguments of an ordinary function. But, hold on, did we not state that functions only receive input through their arguments, without changing them? Well, `two_body_step()` can change the states of the two bodies because Python uses a mechanism that is known as call by object reference. You can easily convince yourself that theses changes persist after calling `two_body_step()`.

> Like functions, class methods can change public as well as protected attributes of their arguments, provided that they are mutable objects (call by object reference).

The gravitational force between the two bodies is computed with the instance method `gravity()`. Inside `two_body_step()`, it is called in the same fashion as `distance()` in the example above (see line 34), with `cls` pointing to the class. As a result, the states of `body1` and `body2` are iterated in the loop through all time steps and successive positions are recorded, like ephemerides, in the arrays `orbit1` and `orbit2`. It is left as an exercise to plot the orbits and to compare them with Fig. 4.8.

Why not split `two_body_step()` into separate updates for the two bodies, which could be implemented as instance methods? In that case, the gravitational force between the bodies would be evaluated twice rather than applying Newton's third law to compute the their accelerations at once. Since force computation involves the most expensive arithmetic operations, this should be avoided. You can try to write a version for three-body interactions and apply it, for instance, to the Algol system (see Sect. 4.3). In systems consisting of many gravitating bodies, such as the stellar cluster discussed in Sect. 4.5, performance becomes the main issue. Direct summation of gravitational forces over all pairs of bodies becomes quickly intractable and approximate algorithms have to be applied. If you want to test it, see Exercise 4.18.

Although defining objects for pieces of data helps to write well structured code, the overhead can become problematic in numerical applications. As a consequence, it is preferable to collect data in objects rather than representing each item individually by an object. You can do an OOP project of your own in Exercise 4.15. The task is to write a class for a large number of test particles orbiting a gravitating central mass.

Appendix B
Making Python Faster

B.1 Using Arrays

The native data structure for ordered collections of data items in Python is a list. At first glance, lists are very similar to NumPy arrays. For example, the days of equinoxes and solstices in 2020 could be defined as a list:

```
1  N = [79, 171, 265, 355]
```

Compare this to the definition of the array N in Sect. 2.1. The list on the right-hand side is just what is passed as argument to `np.array()`. What does this function do? Well, it converts a given list into an array. If you enter

```
2  N[1], type(N[1]), type(N)
```

you will see the output

```
   (171, int, list)
```

So `N[1]` is an integer, which is the second element of the list N (indexing works in the same way as with arrays). Now let us convert this list into an array and assigned it to the same name as before. The original list will be deleted in this case:

```
3  import numpy as np
4
5  # convert to array
6  N = np.array(N)
7
8  print(N[1], N[1].dtype, N.dtype)
```

Now the data type of the element `N[1]` is given by the NumPy attribute `dtype`:

```
   171 int64 int64
```

© Springer Nature Switzerland AG 2021
W. Schmidt and M. Völschow, *Numerical Python in Astronomy and Astrophysics*,
Undergraduate Lecture Notes in Physics,
https://doi.org/10.1007/978-3-030-70347-9

In contrast to a list, this data type is also an attribute of the whole array because the data type must be uniform, i.e. all elements must have an identical data type.

This does not apply to lists:

```
 9   # redefine list
10   N = [79, "summer solstice", 265, "winter solstice"]
11
12   N[1], type(N[1]), type(N)
```

Now some of the elements are strings in place of integers:

```
    ('summer solstice', str, list)
```

You can easily convince yourself that the first and third elements are still of type int. Although the meaning of the expressions in line 10 might be the same for a human, the meaning that pertains to these data types is fundamentally different in Python. What happens if N is again converted into an array? Try for yourself. You are probably up for a surprise.

Uniformity of the data type is important for the memory layout of NumPy arrays. The speed of modern CPUs is fast compared to the time required to load data from the computer's main memory (RAM). If data cannot be loaded more or less in big chunks, the CPU will inevitably be idle for a while, before it receives the next piece of data. This situation is typically encountered when working with Python lists because different elements of the list can be stored at random positions in memory. NumPy arrays, on the other hand, alleviate the problem by arranging elements consecutively in memory.[2] The total memory required to store an array is given by the number of elements times the size of each element in bytes and there is a simple mapping between array elements and their positions in memory.[3]

It turns out that this has a substantial impact on the efficiency of numerical computations using NumPy functions or array operations. As an example, let us compute the Planck spectrum for a given temperature (see Sect. 3.1.2). First, we do this by using lists. For convenience, we use physical constants from scipy.constants.

```
1   import math
2   from scipy.constants import h,c,k,sigma
3
4   # list of wavenumbers
5   n = 1000
6   lambda_max = 2e-6
7   lambda_step = lambda_max/n
8   wavelength = [i*lambda_step for i in range(1,n+1)]
```

---

[2]This is possible through a hierarchy of caches which provide access to blocks of memory at much higher speed than the main memory.

[3]This is also true for multi-dimensional arrays, although the mapping is slightly more complicated.

The last line shows how to iteratively create a Python list of uniformly spaced wavenumbers by means of an inline **for** loop. The following function calculates the intensity for a list of wavelengths. The effective temperature of the Sun is set as default.

```
 9  def planck_spectrum(wavelength, T=5778):
10
11      # create empty list
12      spectrum = []
13
14      # loop through wavelengths and append flux values
15      for val in wavelength:
16          spectrum.append(2*h*c**2 /
17              (val**5 * (math.exp(min(700, h*c/(val*k*T))) - 1)))
18
19      return spectrum
```

Here, we have an explicit loop that runs through all wavelengths and computes the corresponding intensity using the exponential function from the `math` module. The result is appended to the list `spectrum`, which is initialized as an empty list without any elements in line 12. This is an important difference between lists and arrays. The latter always have to be initialized with non-zero length. Moreover, the method `append()` in the example above modifies an existing object. As a result, appended elements can be located anywhere in memory. In contrast, NumPy's `append()` function returns a newly allocated array that is a copy of the original array plus one or more new elements.

A tool for performance measurement that is particularly easy to use is the magic command `timeit` in interactive Python (alternatively use `-m timeit` as command line option if Python is executed on the shell):

```
20  %timeit planck_spectrum(wavelength)
```

which outputs (numbers will differ depending on your computer architecture):

834 ms ± 3.44 ms per loop (mean ± std. dev. of 7 runs, 1000 loops each)

This tells us that execution of `planck_spectrum(wavelength)` took 834 ms (a little less than one second) averaged over 7 runs. Since `timeit` is tailored to small code snippets that are executed in a very short time, the measurement period is artificially increased by a large number of loops for higher precision (the number of loops is automatically adjusted).

It is left as an exercise for you to time the implementation of `planck_spectrum()` based on NumPy from Sect. 3.1.2 (do not forget to convert `wavelength` to an array). We get:

50.3 ms ± 1.28 ms per loop (mean ± std. dev. of 7 runs, 10000 loops each)

With NumPy, the computation is more than ten times faster! We can assign the array returned by `planck_spectrum()` to explore its properties:

```
21  solar = planck_spectrum(wavelength)
22  solar.flags
```

The `flags` attribute gives us the following information (abridged):

```
C_CONTIGUOUS : True
F_CONTIGUOUS : True
```

Indeed, the resulting array is contiguous in memory.[4] You may also check the array `wavelength`.

However, this is not the only reason for higher efficiency. Array operations also avoid function calls inside loops, which are particularly costly. Especially beginners might prefer an explicit loop as it is easy to read. For example, we can apply the trapezoidal rule to integrate the Planck spectrum (see Sect. 3.2.2) using a **for** loop:

```
1   def integr_trapez(f, a, b, n):
2
3       # integration step
4       h = (b - a)/n
5
6       # initialisation
7       tmp = 0.5*f(a)
8
9       # loop through subintervals between a+h and b-h
10      for i in range(1,n):
11          tmp += f(a + i*h)
12
13      tmp += 0.5*f(b)
14
15      return h*tmp
```

Timing the integration of the Planck spectrum,

```
16  %timeit integr_trapez(planck_spectrum, 1e-9, 364.7e-9, 100)
```

yields a mean execution time of

```
474 µs ± 13.4 µs per loop (mean ± std. dev. of 7 runs, 1000 loops each)
```

For comparability, we used the Numpy version of `planck_spectrum()`. Now let us see what we get if the integrator from Sect. 3.2.2 is used:

```
76.7 µs ± 893 ns per loop (mean ± std. dev. of 7 runs, 10000 loops each)
```

Once more, the speed up is palpable. Passing an array as function argument and applying `np.sum()` turns out to be much faster than calling the function for single values inside the body of a **for** loop, as in the code example above. So the lesson

---

[4]The two types are relevant for two-dimensional arrays, where the layout can be row-major as in C or column-major as in Fortran.

learned is to avoid explicit loops and especially function calls inside loops whenever possible. Even in cases where this is not feasible, there are means of improving performance, as shown in the following section.

B.2   Cythonizing Code

As a case study, we consider the Strömgren sphere discussed in Sect. 4.1.1. To solve the initial value problem with the fourth-order Runge-Kutta method (RK4), we iteratively call the Python function `rk4_step()`. As demonstrated in the previous section, function calls inside a loop have a negative impact on performance. We can time the complete RK4 integration by defining a wrapper function:

```
 1   import numpy as np
 2
 3   def solve_stroemgren(r0, dt, n_steps):
 4       t = np.linspace(0, n_steps*dt, n_steps+1)
 5       r = np.zeros(n_steps+1)
 6       r[0] = r0
 7
 8       for n in range(n_steps):
 9           r[n+1] = rk4_step(lambda t, r: (1 - r**3)/(3*r**2),
10                             t[n], r[n], dt)
11
12       return (t,r)
```

This allows us to utilize the `%timeit` command introduced in Sect. B.1:

```
13   %timeit solve_stroemgren(0.01, 1e-3, 10000)
```

The measured execution time (subject to the computer in use) is

```
94.3 ms ± 997 µs per loop (mean ± std. dev. of 7 runs, 10 loops each)
```

which is about a tenth of a second.

If this is so, why do we not implement the RK4 scheme for a specific differential equation rather than using a function for arbitrary initial value problems? Without a generic, reusable implementation, it would be necessary to rewrite the code every time a new problem is to be solved. This is time consuming and prone to errors.

A better solution is to translate the Python code for the RK4 scheme into the compilable language C. A compiler produces machine code that is much faster than code that is executed by the Python interpreter. This is made possible by Cython and requires some preparation.[5] To begin with, we need to put the functions we want to turn into C code in a module with file extension `.pyx`. In our example, this means copying the definition of `rk4_step()` into a file named `stroemgren.pyx` (or

---

[5]For more information, see cython.readthedocs.io/en/latest/index.html.

extract the file from the zip-archive for this section). An important modification to take advantage of Cython is *static typing*. Remember that variables in Python are versatile objects without a fixed data type. For example, you can initially assign a float to a variable, then change it to a string and later to something entirely different. As a consquence, there is no way of telling what kind of data is passed as actual argument to a function before it is called. This is nice and extremely flexible, but not favourable in terms of efficiency. In a low-level language such as C, you need to exactly specify the type of every variable, function argument, and return value. So if you want to translate Python code into C via Cython, you should make data types explicit in your Python source code:

```
1  # excerpt from stroemgren.pyx
2  cpdef double crk4_step(f, double t, double x, double dt):
3
4      cdef double k1 = dt * f(t, x)
5      cdef double k2 = dt * f(t + 0.5*dt, x + 0.5*k1)
6      cdef double k3 = dt * f(t + 0.5*dt, x + 0.5*k2)
7      cdef double k4 = dt * f(t + dt, x + k3)
8
9      return x + (k1 + 2*(k2 + k3) + k4)/6.0
```

The C type `double` corresponds to a floating point number in Python. For local variables you need to use the Cython keyword `cdef` followed by the type. In addition, the function will return a static type if it is declared with `cpdef` instead of `def`. Formal arguments of a function are simply prefixed with the type. You might notice that first argument has no type because `f` is the name of a Python function.

As preparation for using the module `stroemgren.pyx`, we create a small Python script `setup.py`:

```
from setuptools import setup
from Cython.Build import cythonize

setup(
    ext_modules=cythonize("stroemgren.pyx")
)
```

The central directive is `cythonize("stroemgren.pyx")`. It instructs Python to cynthonize the code contained in the module `stroemgren.pyx`. For the next step, you need a C compiler, such as the GNU compiler `gcc`, on your system.[6] The command

```
python setup.py build_ext --inplace
```

---

[6]On Linux and Mac systems you can try to type `gcc -v` on the command line. If the compiler is installed, you will get some information, otherwise an error message will result. Anaconda comes with integrated compiler tools. If there is no compiler on your computer, search the web for available compilers for your system.

executed on the command line instructs Cython to produce a shared object file (also know as shared library).[7] Shared libraries are extensions that can be used by programs when they are executed.

Now we can import and use `crk4_step()` just like any other Python function defined in a module:

```
1  import numpy as np
2  from stroemgren import crk4_step
3
4  def solve_stroemgren(r0, dt, n_steps):
5      t = np.linspace(0, n_steps*dt, n_steps+1)
6      r = np.zeros(n_steps+1)
7      r[0] = r0
8
9      for n in range(n_steps):
10         r[n+1] = crk4_step(lambda t, r: (1 - r**3)/(3*r**2),
11                            t[n], r[n], dt)
12
13     return (t,r)
```

A measurement of the execution time confirms that using `crk4_step()` instead of `rk4_step()` makes a difference:

```
14  %timeit solve_stroemgren(0.01, 1e-3, 10000)
```

The measured execution time (subject to the computer in use) is

21.5 ms ± 1.14 ms per loop (mean ± std. dev. of 7 runs, 10 loops each)

The gain in speed is about a factor four.

It turns out that we can do even better if we refrain from passing a Python function as argument (something that does not translate well into C). In the example above, a Python lambda is used to specify the derivate $d\tilde{r}/d\tilde{t}$ defined by Eq. (4.11). A more efficient computation can be accomplished by using a pure C function in the module `stroemgren.pyx`:

```
15  # excerpt from stroemgren.pyx
16  cdef double rdot(double t, double r):
17      return (1.0 - r**3)/(3.0*r**2)
```

The Cython keyword `cdef` indicates that this function will be compiled as a C function and cannot be called directly by a Python program. For this reason, it has to be referenced explicitly from within the RK4 integrator:

---

[7]If you list the files in your work directory, you will see that also a file named `stroemgren.c` is produced. This file contains the C code generated by Cython from which the C compiler produces the shared object file. It should be mentioned that the intermediate C code is not intended to be read by humans.

```
18    # excerpt from stroemgren.pyx
19    cpdef double stroemgren_step(double t, double r, double dt):
20
21        cdef double k1 = dt * rdot(t, r)
22        cdef double k2 = dt * rdot(t + 0.5*dt, r + 0.5*k1)
23        cdef double k3 = dt * rdot(t + 0.5*dt, r + 0.5*k2)
24        cdef double k4 = dt * rdot(t + dt, r + k3)
25
26        return x + (k1 + 2*(k2 + k3) + k4)/6.0
```

Of course, we give away versatility since the integrator `stroemgren_step()` requires a particular function name. However, we could easily change its definition to solve other first-order differential equations. Let us see how well the fully cythonized functions perform in the following program:

```
1     from stroemgren import stroemgren_step
2
3     def solve_stroemgren(r0, dt, n_steps):
4         t = np.linspace(0, n_steps*dt, n_steps+1)
5         r = np.zeros(n_steps+1)
6         r[0] = r0
7
8         for n in range(n_steps):
9             r[n+1] = stroemgren_step(t[n], r[n], dt)
10
11        return (t,r)
```

Executing

```
12    %timeit solve_stroemgren(0.01, 1e-3, 10000)
```

we finally get

```
7.97 ms ± 260 µs per loop (mean ± std. dev. of 7 runs, 100 loops each)
```

Compared to the plain Python version, we have achieved a speed-up by more than a factor of ten. You might still think that is of no concern whether your Python program finishes in a somewhat shorter or longer fraction of a second. However, if you increase the number of steps to 100.000—the convergence study in Sect. 4.1.1 suggests that this is the recommended number of steps—you will begin to notice the difference without using `timeit`. If it comes to more demanding problems, for example, the colliding disk simulations in Sect. 4.1.1, performance optimization becomes an issue, and even more so in scientific data analysis in research projects.

You need to find a balance between the versatility offered by Python and the efficiency of languages such as C and Fortran that serves your purpose. There is much more to Cython than what we can cover here. We have given you only a first taste and your are invited to explore the capabilities on your own.[8] Since Cython

---

[8]A good starting point are the online tutorials: cython.readthedocs.io/en/latest/src/tutorial.

allows you to call external C functions from Python, an alternative to translating Python into C is to implement critical parts of an algorithm directly as C functions. Moreover, code can be executed very fast on GPUs, as outlined in the following section.

B.3 Parallelization and GPU Offloading

The N-body problem discussed in Sect. 4.5 belongs to the class of N^2 problems. That is, if we double the number N of objects, the computational cost per numerical timestep increases by a factor of $2^2 = 4$. Therefore, simulating large N-body systems becomes prohibitively expensive if the computation is carried out by a single processor core. Since modern CPUs have a multi-core architecture, it is possible to speed up the computationally most demanding part by distributing the workload over several cores. Without going into the gory details, you can think of cores as identical compute units in a processor (CPU). Each core can carry out an independent task. Alternatively, a program can be split into several threads running in parallel on different cores. While each task has its own data in memory, threads may share data. The `threading` and `multiprocessing` modules of the Python Standard Library offer APIs for running Python programs on multiple cores. They are relatively easy to use even for beginners. If you want to give it a try, we recommend the online guide docs.python.org/3/library/concurrency.html. For running fully parallelized codes on high-performance clusters, MPI for Python is available in the `mpi4py` package. However, this is a topic for advanced programming courses.[9]

An alternative is the so-called offloading of computationally intensive tasks to accelerators. The most common accelerators are now Graphics Processing Units (GPU). Especially for machine learning, the speedup can be dramatic. Compared to a CPU, a GPU has a much larger number of execution units and is capable of performing hundreds or even thousands of floating point operations in parallel.[10] The downside is that GPUs have significantly tighter memory restrictions and it is not straightforward to write code that runs efficiently on a GPU. In Python, the library `PyOpenCL` allows you to create special functions called kernels that can be executed on a GPU.[11] For example, the computation of the forces in a N-body simulations can be put into a kernel. For a small number of bodies, copying data such as the positions from the main memory of the computer into the GPU's local

---

[9] See mpi4py.readthedocs.io/en/stable/index.html.

[10] For instance, high-end graphics cards such as the Nvidia Titan V have as many as 5120 individual execution units.

[11] PyOpenCL is the Pythonic flavor of OpenCL (Open Computing Language), a standardized API to run programs on all kinds of accelerators. For further information, see pypi.org/project/pyopencl.

memory is a bottleneck that slows down the code. Beyond roughly 1000 bodies, however, substantial performance gains are possible. If you are interested in GPU programming, you can find more on the subject in specialized textbooks and on the web.

References

1. J.M. Kinder, P. Nelson, *A Student's Guide to Python for Physical Modeling* (Princeton University, Oxford, 2015)
2. J. Freely, *Aladdin's Lamp: How Greek Science Came to Europe Through the Islamic World*, 1st edn. (Alfred A. Knopf, New York, 2009)
3. H. Karttunen, P. Kröger, H. Oja, M. Poutanen, K.J. Donner, *Fundamental Astronomy* (Springer, Berlin, 2017). https://doi.org/10.1007/978-3-662-53045-0
4. B.W. Carroll, D.A. Ostlie, *An Introduction to Modern Astrophysics*, 2nd edn. (Pearson, San Francisco, 2014)
5. P. Goldreich, S. Soter, Icarus **5**(1), 375 (1966). https://doi.org/10.1016/0019-1035(66)90051-0
6. J.B. Holberg, M.A. Barstow, F.C. Bruhweiler, A.M. Cruise, A.J. Penny, Astrophys. J. **497**(2), 935 (1998). https://doi.org/10.10862F305489
7. R.H. Landau, *Computational Physics: Problem Solving with Python* (Wiley-VCH, 2015)
8. P. Bretagnon, G. Francou, Astron. Astrophys. **202**, 309 (1988)
9. J. Lequeux, *The Interstellar Medium* (Springer, Berlin, 2005). https://doi.org/10.1007/b137959
10. W.H. Press, S.A. Teukolsky, W.T. Vetterling, B.P. Flannery, *Numerical Recipes the Art of Scientific Computing*, 3rd edn. (Cambridge University Press, Cambridge, 2007)
11. J.E. Lyne, Nature **375**(6533), 638 (1995). https://doi.org/10.1038/375638a0
12. C.W. Misner, K.S. Thorne, J.A. Wheeler, Gravitation. Freeman (1973)
13. R. Montgomery, Sci. Am. **321**(2), 66 (2019)
14. M.C. Schroeder, N.F. Comins, Astronomy **16**(12), 90 (1988)
15. Planck Collaboration, P. A. R. Ade, N. Aghanim, and 259 more, A&A **594**, A13 (2016). 10.1051/0004-6361/201525830. URL https://doi.org/10.1051/0004-6361/201525830
16. S. Perlmutter, G. Aldering, G. Goldhaber, and 30 more, ApJ **517**(2), 565 (1999). https://doi.org/10.1086/307221
17. I. Goodfellow, Y. Bengio, A. Courville, *Deep Learning* (MIT Press, 2016). http://www.deeplearningbook.org
18. S. Raschka, V. Mirjalili, *Python Machine Learning. Machine Learning and Deep Learning with Python, Scikit-learn, and TensorFlow*, 2nd edn. fourth release, [fully revised and updated] edn. Expert insight (Packt Publishing, 2018)
19. A. Baillard, E. Bertin, V. de Lapparent, P. Fouqué, S. Arnouts, Y. Mellier, R. Pelló, J.-F. Leborgne, P. Prugniel, D. Makarov, L. Makarova, H.J. McCracken, A. Bijaoui, L. Tasca, A&A **532**, A74 (2011). https://doi.org/10.1051/0004-6361/201016423

© Springer Nature Switzerland AG 2021
W. Schmidt and M. Völschow, *Numerical Python in Astronomy and Astrophysics*,
Undergraduate Lecture Notes in Physics,
https://doi.org/10.1007/978-3-030-70347-9

20. N. Piskunov, in *Second BRITE-Constellation Science Conference: Small Satellites - Big Science*, vol. 5, ed. by K. Zwintz, E. Poretti (2017), pp. 209–213
21. J.B. Kaler, *Stars and Their Spectra an Introduction to the Spectral Sequence*, 1st edn. (Cambridge University Press, 1989).

Index

© Springer Nature Switzerland AG 2021
W. Schmidt and M. Völschow, *Numerical Python in Astronomy and Astrophysics*,
Undergraduate Lecture Notes in Physics,
https://doi.org/10.1007/978-3-030-70347-9

Printed in the United States
by Baker & Taylor Publisher Services